ESTIMATOR'S PIPING MAN-HOUR MANUAL

F I F T H

E D I T I O N

Man-Hour Manuals and Other Books by John S. Page

Conceptual Cost Estimating Manual

Cost Estimating Man-Hour Manual
for Pipelines and Marine Structures

Estimator's Electrical Man-Hour Manual/3rd Edition

Estimator's Equipment Installation
Man-Hour Manual/3rd Edition

Estimator's General Construction
Man-Hour Manual/2nd Edition

Estimator's Man-Hour Manual on Heating,
Air Conditioning, Ventilating, and Plumbing/2nd Edition

Estimator's Piping Man-Hour Manual/5th Edition

John S. Page has wide experience in cost and labor estimating, having worked for some of the largest construction firms in the world. He has made and assembled numerous types of estimates including lump-sum, hard-priced, and scope, and has conducted many time and method studies in the field and in fabricating shops. Mr. Page has a B.S. in civil engineering from the University of Arkansas and received the Award of Merit from the American Association of Cost Engineers in recognition of outstanding service and cost engineering.

ESTIMATOR'S PIPING MAN-HOUR MANUAL

FIFTH EDITION

JOHN S. PAGE

Gulf Professional Publishing
An Imprint of Elsevier

Estimator's Piping Man-Hour Manual
Fifth Edition

Permissions may be sought directly from Elsevier's Science and Technology Rights Department in Oxford, UK. Phone: (44) 1865 843830, Fax: (44) 1865 853333, e-mail: permissions@elsevier.co.uk. You may also complete your request on-line via the Elsevier homepage: http://www.elsevier.com by selecting "Customer Support" and then "Obtaining Permissions".

Originally Published by; Gulf Professional Publishing
An Imprint of Elsevier
Houston, TX

Library of Congress Cataloging-in Publication Data

Page, John S.
    Estimator's piping man-hour manual / John S. Page.–5th ed.
        p.   cm.
    Includes bibliographical references and index.
        1. Pipe-fitting–Estimates–United States.   2. Labor time.   I.Title.
    TH6721.P3   1999
    696'.2'0299–dc21   99-18583
                                                                          CIP
ISBN - 13: 978-0-88415-259-0
ISBN - 10: 0-88415-259-6
Printed on acid-free paper (∞)
For information, please contact:
Manager of Special Sales
Butterworth–Heinemann
225 Wildwood Avenue
Woburn, MA 01801-2041
Tel: 781-904-2500
Fax: 781-904-2620
For information on all Butterworth–Heinemann publications available, contact out World Wide Web home page at:
http://www.bh.com

Tranferred To Digital Printing 2011

CONTENTS

Preface, xi
   The Human Factor in Estimating, xi
Introduction, xii

Section One—SHOP FABRICATION OF PIPE AND FITTINGS

Section Introduction 1
Shop Handling Scheduled Pipe for Fabrication 2
Shop Handling Heavy Wall Pipe for Fabrication 3
Shop Handling Large O.D. Pipe for Fabrication 4
Notes On Pipe Bends 5
Standard Types of Bends 6
Pipe Bends—Schedule 20 to 100 Inclusive 7
Pipe Bends—Schedule 120, 140 and 160 8
Pipe Bends—Heavy Wall—45° or Less 9
Pipe Bends—Heavy Wall—Over 45° to 90° Inclusive 10
Pipe Bends—Large O.D. Sizes 11
Attaching Flanges—Screwed Type 12
Attaching Flanges—Screwed Type 13
Attaching Flanges—Screwed Type 14
Attaching Flanges—Slip-On Type 15
Attaching Flanges—Slip-On Type 16
Attaching Flanges—Weld Neck Type 17
Attaching Orifice Flanges—Slip-On and Threaded Types 18
Attaching Orifice Flanges—Weld Neck Type 19
General Welding Notes 20
Butt Welds—Inert Gas Shielded Root Pass 21
Machine Butt Welds 22
Manual Butt Welds—Scheduled 23
Manual Heavy Wall Butt Welds 24
Manual Large O.D. Butt Welds 25
90° Welded Nozzles 26
90° Welded Nozzles—Reinforced 27
Large O.D. 90° Nozzle Welds 28
Large O.D. 90° Nozzle Welds—Reinforced 28
45° Welded Nozzles 29
45° Welded Nozzles—Reinforced 30
Large O.D. 45° Nozzle Welds 31
Large O.D. 45° Nozzle Welds—Reinforced 31
Concentric Swedged Ends 32
Eccentric Swedged Ends 33
End Closures—Pressure Type 34
Heavy Wall End Closure—Pressure Type 35
Large O.D. Pipe End Closures—Pressure Type 36
90° Coupling Welds and Socket Welds 37
'Olet Type Welds 38

Flame Cutting Pipe—Scheduled 39
Flame Cutting Pipe—Heavy Wall 40
Flame Cutting Pipe—Large O.D. Sizes 41
Machine Cutting Pipe—Scheduled 42
Machine Cutting Pipe—Heavy Wall 43
Machine Cutting Pipe—Large O.D. Sizes 44
Flame Beveling Pipe—"V" Type 45
Machine Beveling Pipe—"U" Type, "V" Type and Double Angle 46
Beveling Heavy Wall Pipe 47
Beveling Large O.D. Pipe 48
Threading Pipe—Including Cut 49
Welded Carbon Steel Attachments 50
Drilling Holes in Welded Attachments 50
Machining Inside of Pipe 51
Machining Inside of Large O.D. Pipe 52
Boring Inside Diameter of Pipe and Installing Straightening Vanes 53
Installing Flow Nozzles 54
Preheating Butt Welds and Any Type Flange Welds 55
Preheating Heavy Wall Pipe Butt Welds 56
Preheating Large O.D. Pipe Butt Welds 57
Preheating 90° Nozzle Welds 58
Preheating Large O.D. 90° Nozzle Welds 59
Local Stress Relieving—Scheduled 60
Local Stress Relieving—Heavy Wall 61
Local Stress Relieving—Large O.D. Sizes 62
Full Furnace Stress Relieving and Heating Treatment 63
Radiographic Inspection—Scheduled 64
Radiographic Inspection—Heavy Wall 65
Radiographic Inspection—Large O.D. Sizes 66
Magnetic or Dye Penetrant Inspection of Welded Joints 67
Magnetic or Dye Penetrant Inspection of Welded Joints 68
Testing Fabricated Assemblies—Flanged Ends 69
Testing Fabricated Assemblies—Plain or Beveled Ends 70
Testing Fabricated Assemblies—Heavy Wall 71
Access Holes 72
Miscellaneous Fabrication Operations 73
Man Hours Per Foot of Cylindrical Coil Fabrication Bending Only 74

Section Two—FIELD FABRICATION AND ERECTION

Section Introduction 75
Handling and Erecting Straight Run Pipe—Scheduled 76
Handling and Erecting Straight Run Pipe—Heavy Wall 77
Handling and Erecting Straight Run Pipe—Large O.D. Size 78
Handling and Erecting Fabricated Spool Pieces—Scheduled 79
Handling and Erecting Fabricated Spool Pieces—Heavy Wall 80
Handling and Erecting Fabricated Spool Pieces—Large O.D. Sizes 81
Making on Screwed Fittings and Valves 82
Field Handling Valves 83
Field Erection Bolt-Ups 84
Attaching Flanges—Screwed Type 85
Attaching Flanges—Screwed Type 86

vi

Attaching Flanges—Screwed Type 87
Attaching Flanges—Slip-On Type 88
Attaching Flanges—Weld Neck Type 89
Attaching Orifice Flanges—Slip-On and Threaded Types 90
Attaching Orifice Flanges—Weld Neck Type 91
General Welding Notes 92
Manual Butt Welds—Schedule 93
Manual Butt Welds—Heavy Wall 94
Manual Butt Welds—Large O.D. Sizes 95
90° Welded Nozzles 96
90° Welded Nozzles—Reinforced 97
Large O.D. 90° Nozzle Welds 98
Large O.D. 90° Nozzle Welds—Reinforced 98
45° Welded Nozzles 99
45° Welded Nozzles—Reinforced 100
Large O.D. 45° Nozzle Welds 101
Large O.D. 45° Nozzle Welds—Reinforced 101
Concentric Swedged Ends 102
Eccentric Swedged Ends 103
End Closures—Pressure Type 104
Heavy Wall End Closures—Pressure Type 105
Large O.D. Pipe End Closures—Pressure Type 106
90° Coupling Welds and Socket Welds 107
'Olet Type Welds 108
Flame Cutting Pipe—Scheduled 109
Flame Cutting Pipe—Heavy Wall 110
Flame Cutting Pipe—Large O.D. Sizes 111
Flame Beveling Pipe—"V" Type 112
Flame Beveling Pipe—Large O.D. Sizes 113
Threading Pipe—Including Cut 114
Welded Carbon Steel Attachments 115
Drilling Holes in Welded Attachments 116
Machining Inside of Pipe 117
Machining Inside of Large O.D. Pipe 118
Boring Inside Diameter of Pipe and Installing Straightening Vanes 119
Installing Flow Nozzles—Holding Ring Type 120
Preheating Butt Welds and Any Type Flange Welds 121
Preheating Heavy Wall Pipe Butt Welds 122
Preheating Large O.D. Pipe Butt Welds 123
Preheating 90° Nozzle Welds 124
Preheating Large O.D. 90° Nozzle Welds 125
Local Stress Relieving—Scheduled 126
Local Stress Relieving—Heavy Wall 127
Local Stress Relieving—Large O.D. Sizes 128
Radiographic Inspection—Scheduled 129
Radiographic Inspection—Heavy Wall 130
Radiographic Inspection—Large O.D. Sizes 131
Hydrostatic Testing—Scheduled 132
Hydrostatic Testing—Heavy Wall 133
Hydrostatic Testing—Large O.D. Sizes 134
Access Holes 135
Instrument and Control Piping 136
Soldered Non-Ferrous Fittings 136

PVC-Plastic Pipe 137
Saran Lined Steel Pipe and Fittings 138
Schedule 30 or 40 Rubber-Lined Steel Pipe and Fittings 139
Schedule 40 Lead Lined Steel Pipe and Fittings 140
Flanged Cast Iron Cement Lined Pipe and Fittings 141
Schedule 40 Cement Lined Carbon Steel Pipe with Standard Fittings 142
Double Tough Pyrex Pipe and Fittings 143
Overhead Transite Pressure Pipe—Class 150 144

Section Three—ALLOY AND NON-FERROUS FABRICATION

Section Introduction 145
Shop Handling Pipe for Fabrication 146
Handle and Erect Fabricated Spool Pieces 147
Handle and Erect Straight Run Pipe 148
Pipe Bends 149
Attaching Flanges 150
Make-Ons through 12-in. Handle Valves through 42-in. 151
Field Erection Bolt-Ups 153
All Welded Fabrication 154
Flame Cutting or Beveling 155
Machine Cutting and Beveling Pipe 156
Threading Pipe 157
Welded Attachments and Drilling Holes in Welded Attachments 158
Local Stress Relieving 159
Radiographic Inspection 160
Magnetic or Dye Penetrant Inspection 161
Hydrostatic Testing 162
Access Holes 163

Section Four—PNEUMATIC MECHANICAL INSTRUMENTATION

Section Introduction 164
Liquid Level Gauge Glasses—Transparent Type 165
Liquid Level Gauge Glasses—Transparent Type 166
Liquid Level Gauge Glasses—Transparent Type 167
Liquid Level Gauge Glasses—Transparent Type 168
Liquid Level Gauge Glasses—Reflex Type 169
Liquid Level Gauge Glasses—Reflex Type 170
Liquid Level Gauge Glasses—Reflex Type 171
Liquid Level Gauge Glasses—Reflex Type 172
Pressure Gauges 173
Pneumatic Liquid Level Instruments—Local Mounted 174
Pneumatic Liquid Level Instruments—Local Mounted 175
Pneumatic Pressure Instruments—Local Mounted 176
Pneumatic Temperature Instruments—Local Mounted 177
Thermometers and Thermowells 178
Thermometers and Thermowells 179

Thermowells and Thermocouples 180
Relief Valves—Screwed 181
Relief Valves—Flanged 182
Relief Valves—Flanged 183
Pneumatic Flow Transmitters 184
Flow Indicating Transmitters, Flow Recorders and Flow Controllers 185
Pneumatic Liquid Level Transmitters 186
Control Panel Installation 187
Connecting Pneumatic Panel Board Instruments 188
Connecting Pneumatic Panel Board Instruments 189

Section Five—UNDERGROUND PIPING

Section Introduction 190
Machine Excavation 191
Hand Excavation 192
Rock Excavation 193
Shoring and Bracing Trenches 193
Disposal of Excavated Material 194
Backfilling and Tamping 195
Underground 150 Lbs. B. & S. Cast Iron Pipe 195
Underground Vitrified Clay and Concrete Pipe 196
Socket Clamps for Cast Iron Pipe 197
Pipe Coated with Tar and Field Wrapped by Machine 197

Section Six—HANGERS AND SUPPORTS

Section Introduction 198
Hangers and Supports 199

Section Seven—PAINTING

Section Introduction 200
Surface Area of Pipe for Painting 201
Sand Blast and Paint Pipe 202

Section Eight—PATENT SCAFFOLDING

Section Introduction 203
Erect and Dismantle 204

Section Nine—INSULATION

Section Introduction 205
Indoor Thermal Type 206
Insulation—Hot Pipe 207

Section Ten—SAMPLE ESTIMATE

Section Introduction 208
Sample Job Estimate Form 209
Shop Fabrication—Carbon Steel 210
Shop Fabrication—Alloy 211
Field Erect—Shop Fabricated Piping 211
Field Fabricate and Erect—Screwed 212
Field Fabricate and Erect—Welded 213
Erect Valves—Screwed and Flanged 214
Hangers and Supports 215
Sandblast and Paint Pipe 215
Insulation 215
Hand Excavate 215
Underground Piping 215
Estimate Summary 216

Section Eleven—TECHNICAL INFORMATION

Section Introduction 217
Circumference of Pipe for Computing Welding Material 218
Circumference of Pipe for Computing Welding Material—Heavy Wall 219
Circumference of Pipe for Computing Welding Material—Large O.D. Sizes 220
Weights of Piping Materials—General Notes 221
Weights of Piping Materials—1″ 222
Weights of Piping Materials—1¼″ 223
Weights of Piping Materials—1½″ 224
Weights of Piping Materials—2″ 225
Weights of Piping Materials—2½″ 226
Weights of Piping Materials—3″ 227
Weights of Piping Materials—3½″ 228
Weights of Piping Materials—4″ 229
Weights of Piping Materials—5″ 230
Weights of Piping Materials—6″ 231
Weights of Piping Materials—8″ 232
Weights of Piping Materials—10″ 233
Weights of Piping Materials—12″ 234
Weights of Piping Materials—14″ 235
Weights of Piping Materials—16″ 236
Weights of Piping Materials—18″ 237
Weights of Piping Materials—20″ 238
Weights of Piping Materials—24″ 239
Hanger Load Calculations—General Notes 240
Hanger Diagram 241
Table of Weights 241
Hanger Load Calculations 241
Minutes to Decimal Hours—Conversion Table 248

PREFACE

Updated with the addition of 26 new tables on pneumatic mechanical instrumentation, this fifth edition is written for the majority of estimators who have not had the advantages of years of experience and/or of being associated with a firm that spends thousands of dollars for time studies and research analyses. I believe that the book will decrease the chance of errors and help the partially experienced estimator to determine more accurately the actual direct labor cost for the complete fabrication and installation of process piping for a given industrial or chemical plant.

This book is strictly for estimating direct labor in man hours only. You will not find any costs for materials, equipment usage, warehousing and storing, fabricating, shop set-up, or overhead. These costs can be readily obtained by a good estimator who can visualize and consider job schedule, size, and location. If a material take-off is available, this cost can be obtained from vendors who will furnish the materials. These items must be considered for each individual job.

The following direct man hours (or in the case of alloy and nonferrous materials, the percentages) were determined by gathering hundreds of time and method studies coupled with actual cost of various operations, both in the shop and field on many piping jobs located throughout the country, ranging in cost from $1,000,000 to $5,000,000. By carefully analyzing these many reports, I established an average productivity rate of 70%. The man hours or percentages compiled throughout this manual are based on this percentage.

I wish to call your attention to the introduction on the following pages entitled "Production and Composite Rate," which is the key to this method of estimating.

The Human Factor in Estimating

In this high-tech world of sophisticated software packages, including several for labor and cost estimating, you might wonder what a collection of man-hour tables offers that a computer program does not. The answer is the *human factor*. In preparing a complete estimate for a refinery, petrochemical, or other heavy industrial project one often confronts 12–18 major accounts, and each account has 5–100 or more sub-accounts, depending on the project and its engineering design. While it would seem that such numerous variables provide the perfect opportunity for computerized algorithmic solution, accurate, cost-effective, realistic estimating is still largely a function of human insight and expertise. Each project has unique aspects that still require the seasoned consideration of an experienced professional, such as general economy, projects supervision, labor relations, job conditions, construction equipment, and weather, to name a few.

Computers are wonderful tools. They can solve problems as no human can, but I do not believe construction estimating is their forté. I have reviewed several construction estimating software packages and have yet to find one that I would completely rely on. Construction estimating is an art, a science, and a craft, and I recommend that it be done by those who understand and appreciate all three of these facets. This manual is intended for those individuals.

John Page
Houston, Texas

INTRODUCTION
Production and Composite Rate

This is the golden key that unlocks the gate to the wealth of process pipe estimating information that follows. The most important area to be considered before calculating labor dollars is productivity efficiency. This is a must if the many man-hour tables that follow are to be correctly applied. Productivity efficiency in conjunction with the production elements must be considered for each individual project.

I have found after comparing many projects that production percentages can be classified into five categories and the production elements can be grouped into six different classifications. The six different classes of production elements are:

1. General economy
2. Project supervision
3. Labor relations
4. Job conditions
5. Equipment
6. Weather

The five ranges of productivity efficiency percentages are:

Type	Percentage Range
1. Very Low	10–40
2. Low	41–60
3. Average	61–80
4. Very Good	81–90
5. Excellent	91–100

Although you may agree with the ranges described here, you may still wonder with such a wide percentage range how to determine a definite percentage. To illustrate how simply this is done we will evaluate each of the six elements and give an example with each.

1. GENERAL ECONOMY

This is simply the state of the nation or area in which your project is to be developed. Things that should be evaluated under this category are:

a. Business trends and outlooks
b. Construction volume
c. The employment situation

Let us say that you find these items to be very good or excellent. This may sound good, but actually it means your productivity range will be very low. This is because when business is good, the type of supervision and craftsmen that you will have to draw from will be very poor. This will tend to create bad labor relations between your company and supervision and thus produce unfavorable job conditions. On the other hand if you find the general economy to be of a fairly good average, the productivity efficiency will tend to rise. Under normal conditions there are enough good supervisors and craftsmen to go around and everyone is satisfied, thus creating good job conditions.

Example: To show how to arrive at a final productivity efficiency percentage, let us say we find this element to be of a high average in the area of the project. Since it is of a high average, but by no means excellent, we estimate our productivity percentage at 75%.

2. PROJECT SUPERVISION

What is the caliber of your supervision? What experience have they had? What can you afford to pay them? What have you to draw from? Areas to be looked at under this element are:

a. Experience
b. Supply
c. Pay

Like *general economy* this too must be carefully analyzed. If business is excellent, the chances are that you will have a poor lot to draw from. If business is normal, you will have a fair chance of obtaining good supervision. The contractor who tries to cut overhead by using cheap supervision usually winds up doing a very poor job. This usually results in a dissatisfied client, a loss of profit, and a loss of future work. However, the estimator has no control over this. It must be left to management. All the estimator can do is estimate his projects accordingly.

Example: After careful analysis of the three items listed under this element, we find that our supervision will be normal for this type of work and we arrive at an estimated productivity rate of 70%.

3. LABOR RELATIONS

Have you a good labor relations man in your organization? Are the craftsmen in the area experienced and satisfied? Are there adequate first-class craftsmen in the area? Like project supervision things that should be analyzed under this element are:

a. Experience
b. Supply
c. Pay

The area where your project is to be constructed should be checked to see if the proper experienced craftsmen are available locally or if you will have to rely on travelers to fill your needs. Can and will your organization pay the prevailing wage rates?

Example: Let us say that for a project in a given area we have found our labor relations to be fair but feel that they could be a little better. Since this is the case, we arrive at an efficiency rating of 65% for this element.

4. JOB CONDITIONS

What is the scope of the work and just what is involved in the job? Is the schedule tight or do you have ample time to complete the project? What is the condition of the site? Is it high and dry and easy to drain or is it low and muddy and hard to drain? Will you be working around a plant already in production? Will there be tie-ins making it necessary to shut down various systems of the plant? What will be the relationship between production personnel and construction personnel? Will most of your operations be manual or mechanized? What kind of material procurement will you have? There are many items that could be considered here, dependent on the project; however, we feel that the most important of these items that should be analyzed under this element are as follows:

    a. Scope of work
    b. Site conditions
    c. Material procurement
    d. Manual and mechanized operations

By careful study and analysis of the plans and specifications coupled with a site visitation you should be able to correctly estimate a productivity efficiency percentage for this item.

Example: Let us say that the project we are estimating is a completely new plant and that we have ample time to complete the project but the site location is low and muddy. Therefore, after evaluation we estimate a productivity rating of only 60%.

5. EQUIPMENT

Do you have ample equipment to do your job? What kind of shape is it in? Will you have good maintenance and repair help? The main items to study under this element are:

    a. Usability
    b. Condition
    c. Maintenance and repair

This should be the simplest of all elements to analyze. Every estimator should know what type and kind of equipment his company has as well as what kind of mechanical shape it is in.

Example: Let us assume that our company equipment is in very good shape, that we have an ample supply to draw from, and that we have average mechanics. Since this is the case we estimate a productivity percentage of 70%.

6. WEATHER

Check the past weather conditions for the area in which your project is to be located. During the months that you will be constructing what are the weather predictions based on these past reports? Will there be much rain or snow? Will it be hot and mucky or cold and damp? The main items to check and analyze here are as follows:

        a. Past weather reports
        b. Rain or snow
        c. Hot or cold

This is one of the worst of all elements to be considered. At best, all you have is a guess. However, by giving due consideration to the items as outlined under this element your guess will at least be based on past occurrences.

Example: Let us assume that the weather is about half good and half bad during the period that our project is to be constructed. We must then assume a productivity range of 50% for this element.

We have now considered and analyzed all six elements and in the examples for each individual element have arrived at a productivity efficiency percentage. Let us now group these percentages together and arrive at a total percentage:

Item	Productivity Percentage
1. General economy	75
2. Project supervision	70
3. Labor relations	65
4. Job conditions	60
5. Equipment	70
6. Weather	50
Total	390

Since there are six elements involved, we must now divide the total percentage by the number of elements to arrive at an average percentage of productivity.

$$390 \div 6 = 65\% \text{ average productivity efficiency}$$

At this point we must caution the estimator. This example is only a guide to show a method of arriving at a productivity percentage. By considering the preceding elements for each individual project along with the proper man-hour tables that follow, you can make a good labor value estimate for any place in the world at any time.

Next, we must consider the *composite rate* to correctly arrive at a total direct labor cost, using the man-hours in the following tables.

Most organizations consider the cost of field personnel with a rating of superintendent or greater to be a part of job overhead and that of general foreman or lower as direct job labor cost. The direct man hours on the following pages have been determined on this basis. Therefore, a composite rate should be used when converting the man-hours to direct labor dollars.

The estimator must also again consider labor conditions in the area where the project is to be located. He must determine how many men he will be allowed to use in a crew plus how many crews he will need.

Example: This will illustrate how to obtain a composite rate:

We assume that a certain pipe project will need four 10-man crews and that only one general foreman will be needed to head the four crews.

Rate of pipefitter craft in a given area:
General foreman .... $23.75 per hour
Foreman .... $23.50 per hour
Journeyman .... $23.00 per hour
Fifth-year apprentice .... $18.00 per hour

NOTE: General foreman and foreman are dead weight because they do not work with their tools; however, they must be considered and charged to the composite crew.

Crew for composite rate:
One general foreman...................... 2 hours @ $23.75 = $ 47.50
One foreman .... 8 hours @ $23.50 =   188.00
Nine journeymen......................... 8 hours @ $23.00 =   1,656.00
Fifth-year apprentice.................... 8 hours @ $18.00 =   144.00

Total for 80 hours $2,035.50

$2,035.50 ÷ 80 = $25.44 composite man-hour rate for 100% time.

Note that the man hours are based on an average productivity of 70%. Therefore, the composite rate of $25.44 as figured becomes equal to 70%.

Let us assume that you have evaluated your job and find it to be of a low average with a productivity rating of only 65%. This means a loss of 5% of time paid per man hours. Therefore, your composite rate should have an adjustment of 5% as follows:

$25.44 × 105% = $26.71 composite rate of 65% productivity

Simply by multiplying the number of man hours estimated by the calculated composite rate, you can arrive at a total estimated direct labor cost, in dollar value, for pipe fabrication and installation.

The foregoing explanation should enable the ordinary piping estimator to turn out a better labor estimate and will eliminate much guesswork.

Section One

SHOP FABRICATION OF PIPE AND FITTINGS

It is the intent and express purpose of this section to cover as nearly as possible all operations which may be encountered in a shop engaged in the prefabrication of process piping for any type of industrial or chemical plant.

The man hours listed for the various operations are for labor only, and have no bearing on materials which must be added in all cases for a complete labor and material estimate.

All labor for unloading from railroad cars on trucks, storing in fabrication yard or warehouse, hauling to fabrication area, fabricating and returning to storage area or loading for delivery to erection site have been given due consideration in the man hours listed. *No consideration has been given to overhead or profit in any way.*

For alloy and non-ferrous fabrication, apply the percentages which appear under Section Three to the following pages listing the various shop fabrication operations.

1

SHOP HANDLING PIPE FOR FABRICATION

Carbon Steel Material

Wall Thickness Through Schedule 160

DIRECT MAN HOURS PER FOOT

Pipe Size Inches		Schedule 10 to 60		Schedule 80 to 100		Schedule 120 to 160
1/4	. . . .	0.029	. . . .	0.031	. . . .	0.033
3/8	. . . .	0.029	. . . .	0.031	. . . .	0.035
1/2	. . . .	0.030	. . . .	0.033	. . . .	0.036
3/4	. . . .	0.030	. . . .	0.034	. . . .	0.039
1	. . . .	0.031	. . . .	0.036	. . . .	0.041
1-1/4	. . . .	0.033	. . . .	0.039	. . . .	0.044
1-1/2	. . . .	0.035	. . . .	0.041	. . . .	0.049
2	. . . .	0.036	. . . .	0.044	. . . .	0.053
2-1/2	. . . .	0.039	. . . .	0.048	. . . .	0.059
3	. . . .	0.041	. . . .	0.053	. . . .	0.065
3-1/2	. . . .	0.044	. . . .	0.055	. . . .	0.068
4	. . . .	0.045	. . . .	0.058	. . . .	0.071
5	. . . .	0.048	. . . .	0.063	. . . .	0.079
6	. . . .	0.051	. . . .	0.070	. . . .	0.091
8	. . . .	0.063	. . . .	0.088	. . . .	0.119
10	. . . .	0.079	. . . .	0.110	. . . .	0.149
12	. . . .	0.096	. . . .	0.134	. . . .	0.183
14 OD	. . . .	0.116	. . . .	0.159	. . . .	0.218
16 OD	. . . .	0.138	. . . .	0.186	. . . .	0.254
18 OD	. . . .	0.161	. . . .	0.214	. . . .	0.291
20 OD	. . . .	0.189	. . . .	0.241	. . . .	0.329
24 OD	. . . .	0.210	. . . .	0.273	. . . .	0.370

Man hours include unloading pipe from railroad cars or trucks and placing in shop storage, procuring necessary pipe and materials to fabricate spool piece, transporting necessary materials to point of fabrication and the transporting of finished work to temporary storage.

Units apply to any length spool piece or segment of work.

SHOP HANDLING PIPE FOR FABRICATION

Heavy Wall Carbon Steel Material

DIRECT MAN HOURS PER FOOT

Nominal Pipe Size	WALL THICKNESS IN INCHES							
	.750	1.00	1.25	1.50	1.75	2.00	2.25	2.50
3" or less	0.065	0.078	--	--	--	--	--	--
4	0.071	0.087	0.103	0.119	--	--	--	--
5	--	0.099	0.119	0.139	0.159	0.179	--	--
6	--	0.110	0.129	0.148	0.199	0.218	0.237	--
8	--	0.129	0.146	0.174	0.201	0.228	0.255	0.286
10	--	--	0.149	0.182	0.215	0.248	0.281	0.314
12	--	--	--	0.214	0.245	0.276	0.308	0.339
14	--	--	--	0.246	0.274	0.302	0.330	0.359
16	--	--	--	--	0.284	0.315	0.345	0.377
18	--	--	--	--	--	0.324	0.358	0.392
20	--	--	--	--	--	0.329	0.362	0.405
22	--	--	--	--	--	--	--	0.416
24	--	--	--	--	--	--	--	0.486
	2.75	3.00	3.25	3.50	3.75	4.00	4.25	4.50
10	0.361	0.408	--	--	--	--	--	--
12	0.390	0.441	0.498	0.563	--	--	--	--
14	0.413	0.467	0.528	0.597	0.669	0.743	--	--
16	0.434	0.490	0.554	0.626	0.701	0.778	--	--
18	0.451	0.501	0.576	0.651	0.729	0.809	--	--
20	0.466	0.527	0.596	0.673	0.754	0.837	0.921	1.022
22	0.478	0.540	0.610	0.689	0.772	0.857	0.951	1.046
24	0.494	0.558	0.631	0.713	0.798	0.886	0.983	1.081
	4.75	5.00	5.25	5.50	5.75	6.00		
20	1.124	1.225	1.323	1.429	1.529	1.636		
22	1.151	1.255	1.355	1.463	1.565	1.675		
24	1.189	1.296	1.400	1.512	1.618	1.731		

Man hours include unloading pipe from railroad cars or trucks and placing in shop storage, procuring necessary pipe and materials to fabricate spool piece, transporting necessary materials to point of fabrication and the transporting of finished work to temporary storage.

Units apply to any length spool piece or segment of work.

SHOP HANDLING PIPE FOR FABRICATION

Large O.D. Sizes Carbon Steel Material

DIRECT MAN HOURS PER FOOT

O. D. PIPE INCHES	WALL THICKNESS IN INCHES							
	.500 Or Less	.750	1.00	1.25	1.50	1.75	2.00	2.25
26	0.222	0.234	0.270	0.285	0.303	0.360	0.376	0.410
28	0.251	0.264	0.290	0.322	0.352	0.380	0.410	0.440
30	0.268	0.282	0.307	0.338	0.363	0.403	0.451	0.490
32	0.286	0.301	0.334	0.364	0.402	0.434	0.469	0.502
34	0.304	0.320	0.352	0.387	0.427	0.461	0.502	0.534
36	0.336	0.354	0.386	0.421	0.453	0.503	0.563	0.610
38	0.353	0.372	0.407	0.445	0.479	0.532	0.593	0.642
40	0.372	0.392	0.428	0.468	0.504	0.560	0.624	0.676
42	0.405	0.426	0.464	0.506	0.543	0.603	0.675	--
44	0.422	0.444	0.484	0.528	0.568	0.634	0.708	--
46	0.442	0.465	0.506	0.552	0.593	0.662	0.741	--
48	0.473	0.498	0.543	0.592	0.633	0.703	0.780	--
54	0.542	0.570	0.621	0.677	0.723	0.803	0.891	--
60	0.610	0.642	0.700	0.763	0.813	0.902	1.001	--
	2.50	2.75	3.00	3.25	3.50	3.75	4.00	4.25
26	0.446	0.514	0.570	0.640	0.720	0.808	0.900	0.993
28	0.453	0.521	0.580	0.650	0.726	0.819	0.910	1.000
30	0.505	0.530	0.588	0.658	0.734	0.830	0.927	1.020
32	0.517	0.543	0.595	0.670	0.740	0.842	0.940	1.041
34	0.550	0.578	0.607	0.690	0.750	0.860	0.954	1.073
36	0.628	0.659	0.692	0.734	0.778	0.877	0.970	1.105
	4.50	4.75	5.00	5.25	5.50	5.75	6.00	
26	1.100	1.194	1.305	1.410	1.520	1.630	1.751	
28	1.120	1.244	1.370	1.453	1.570	1.695	1.810	
30	1.135	1.299	1.440	1.480	1.600	1.765	1.880	
32	1.170	1.359	1.480	1.520	1.660	1.830	1.960	
34	1.204	1.400	1.520	1.570	1.690	1.855	1.995	
36	1.233	1.480	1.560	1.600	1.740	1.895	2.100	

Man hours include unloading pipe from railroad cars or trucks and placing in shop storage, procuring necessary pipe and materials to fabricate spool piece, transporting necessary materials to point of fabrication and the transporting of finished work to temporary storage.

Units apply to any length spool piece or segment of work.

NOTES ON PIPE BENDS

Minimum Bending Radii: Man hours shown for pipe bends are based upon a minimum bending radii of 5 nominal pipe size diameters, with the exception of large sizes and/or lighter walls which must be bent on longer radii. For bends having a radius of less than 5 diameters add 50% to man hours shown.

Welding Long Bends: When it is necessary to weld together two or more pieces of pipe to produce the length required in the pipe bend, add the man hours for welding.

Compound Bends: Man hours of pipe bends other than the standard types illustrated or with bends in more than one plane are obtained by adding together the man hours of the component bends that are combined to produce the compound bend.

Bends Without Tangents: For Pipe Bends (Sch. 160 and less) ordered without tangents, add 15% to man hours shown.

Bends With Long Arcs: For pipe bending with an arc exceeding 10 feet; add 100% to the bending man hours shown for each additional 10 feet of arc or part thereof.

Connecting Tangents: No. 5 Offset Bends and No. 7 U-Bends are to be considered as such only when the bends are continuous arcs or if the tangent between arcs of the same radius is 1'-0" long or less. If the tangent between arcs is longer than 1'-0", the bends should be considered as compound bends, *i. e.*, double angle bends, double quarter bends, etc.

No. 9 Expansion U-Bends are to be considered as such only when the bends are continuous arcs or if the tangents between the "U" and the 90° bends, of the same radius are 1'-0" or less. When such tangents are longer than 1'-0" the bends should be man houred as one "U" and two 90° Pipe Bends. Bends from 181° to 359° should be man houred at the same man hours as a No. 11 Bend.

Offset Bends: No. 5 Offset Bends are considered as such only when each angle is 90° or less and the connecting tangents between arcs are within the maximum of 1'-0" specified in the preceding note.

Beveled Ends: If Pipe Bends are to have the ends beveled for welding add the man hours for beveling.

Thread Ends: If Pipe Bends are to have the ends threaded only, add the man hours for threading.

Flanged Ends: If Pipe Bends are to have the ends fitted with screwed flanges, slip-on weld flanges, welding neck flanges, or lap joints add the man hours applicable for this operation.

Preparation For Intermediate Field Welds: When Pipe Bends, particularly No. 9, 10 or 11, are too bulky for transporting or handling and therefore, must be furnished in two or more sections for assembly in the field, an extra charge should be made for the additional cuts and bevels.

Unlisted Sizes: For unlisted sizes, use the man hours of the next larger shown size.

STANDARD TYPES OF BENDS

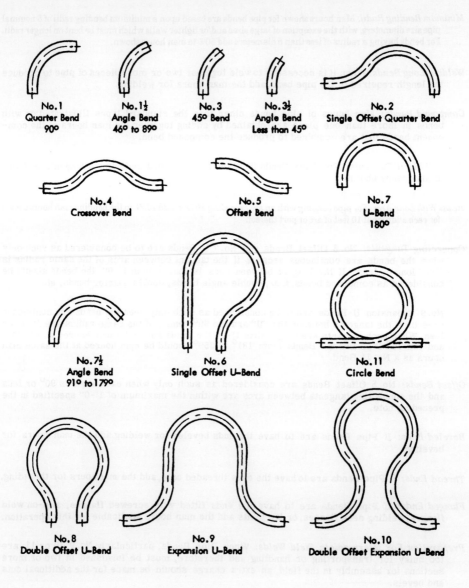

No. 1
Quarter Bend
90°

No. 1½
Angle Bend
46° to 89°

No. 3
45° Bend

No. 3½
Angle Bend
Less than 45°

No. 2
Single Offset Quarter Bend

No. 4
Crossover Bend

No. 5
Offset Bend

No. 7
U–Bend
180°

No. 7½
Angle Bend
91° to 179°

No. 6
Single Offset U–Bend

No. 11
Circle Bend

No. 8
Double Offset U–Bend

No. 9
Expansion U–Bend

No. 10
Double Offset Expansion U–Bend

PIPE BENDS

Schedule Numbers 20 to 100 Inclusive

Labor Only for Making Pipe Bends with Plain Ends

Carbon Steel Material
Double Extra Strong Weight

NET MAN HOURS EACH

Size Ins.	No.1-90° No.1-1/2-46° to 89° No.3-45° No.3-1/2-Less than 45°	No.7 U-180° No.7-1/2-91° to 179°	No. 5 Offset	No.2 Single Offset Quarter	No.6 Single Offset "U"	No.4 Cross-over No.9 Exp. "U"	No.8 Double Offset "U"	No.10 Double Offset Exp. No.11 Circle
1	1.6	2.0	2.4	3.0	3.6	3.5	4.4	4.9
1-1/4	1.8	2.3	2.8	3.4	4.5	4.6	5.2	5.7
1-1/2	2.0	2.6	3.1	3.8	5.0	5.2	5.9	6.6
2	2.3	3.0	3.6	4.6	5.5	6.0	6.0	8.1
2-1/2	2.9	4.1	4.5	5.8	7.2	7.5	8.9	11.7
3	3.1	4.7	5.2	6.3	8.3	8.9	10.0	12.8
3-1/2	3.9	5.2	5.9	7.8	9.7	10.4	12.4	15.8
4	4.4	6.3	6.9	8.6	10.8	11.7	13.7	17.3
5	5.9	7.8	8.3	11.0	13.6	14.6	17.4	22.0
6	7.2	9.2	10.1	12.8	16.0	17.4	21.9	25.8
8	9.4	12.6	12.8	17.3	23.5	25.0	27.3	33.0
10	12.7	17.4	17.4	22.8	31.7	33.8	38.4	43.9
12	17.3	24.4	23.6	32.1	44.8	45.6	61.1	69.4
14 OD	22.0	35.4	31.4	42.8	67.1	67.1	88.6	103.4
16 OD	28.7	54.4	41.3	56.9	83.2	83.2	--	--
18 OD	37.0	72.0	52.3	--	--	--	--	--
20 OD	48.0	--	66.8	--	--	--	--	--
24 OD	81.2	--	103.4	--	--	--	--	--

For General Notes on pipe bends, see pages 5 and 6.

PIPE BENDS

Schedule Numbers 120, 140 and 160

Labor Only for Making Pipe Bends with Plain Ends

Carbon Steel Material
Double Extra Strong Weight

NET MAN HOURS EACH

Size Ins.	No.1-90° No.1-1/2-46° to 89° No.3-45° No.3-1/2-Less than 45°	No.7 U-180° No.7-1/2-91° to 179°	No.5 Offset	No.2 Single Offset Quarter	No.6 Single Offset "U"	No.4 Cross-over No.9 Exp."U"	No.8 Double Offset "U"	No.10 Double Offset Exp. No.11 Circle
1	2.0	2.3	2.8	3.6	4.0	4.7	5.4	6.2
1-1/4	2.1	2.6	3.2	4.2	5.2	5.5	6.5	7.2
1-1/2	2.4	3.1	3.8	4.6	6.2	6.5	7.3	8.6
2	2.7	3.6	4.4	5.4	7.0	7.5	8.1	9.4
2-1/2	3.4	4.7	5.7	6.9	8.6	9.2	10.3	11.7
3	3.9	5.5	6.1	7.7	9.7	10.4	11.7	14.6
4	5.5	7.3	8.5	10.1	13.4	14.3	15.5	18.9
5	6.0	9.2	10.0	12.8	16.4	17.4	20.3	25.7
6	8.5	11.0	11.8	15.5	19.2	20.1	24.8	32.1
8	11.4	14.8	15.2	20.3	27.5	29.3	32.1	40.3
10	14.8	20.8	20.9	26.7	37.0	39.3	45.4	59.5
12	20.3	28.3	28.8	36.6	53.0	54.9	65.8	87.8
14 OD	26.6	45.8	39.7	50.9	75.8	75.8	--	--
16 OD	34.4	64.6	50.9	66.4	103.4	103.4	--	--
18 OD	44.0	82.7	64.7	--	--	--	--	--
20 OD	54.4	--	--	--	--	--	--	--
24 OD	89.2	--	--	--	--	--	--	--

For General Notes on pipe bends, see pages 5 and 6.

PIPE BENDS

Heavy Wall—45° or Less

Labor only for Making Pipe Bends With Plain Ends
Carbon Steel Material

NET MAN HOURS EACH

Nominal Pipe Size	WALL THICKNESS IN INCHES							
	7.50	1.00	1.25	1.50	1.75	2.00	2.25	2.50
4″ or less	6.6	7.5	8.7	10.1	--	--	--	--
5	--	8.6	10.1	11.8	13.5	15.7	--	--
6	--	10.2	11.8	14.2	16.0	19.4	22.6	--
8	--	13.9	14.4	17.0	19.7	23.0	27.2	31.4
10	--	--	17.8	19.5	21.5	24.8	28.7	33.8
12	--	--	--	22.1	23.3	26.9	31.5	36.8
14	--	--	--	25.4	26.4	29.4	34.6	40.1
16	--	--	--	--	32.4	35.3	39.2	45.8
18	--	--	--	--	--	40.3	45.7	53.3
20	--	--	--	--	--	51.8	52.2	62.0
22	--	--	--	--	--	--	--	69.8
24	--	--	--	--	--	--	--	89.2
	2.75	3.00	3.25	3.50	3.75	4.00	4.25	4.50
10	34.9	38.2	--	--	--	--	--	--
12	39.7	43.9	47.2	51.6	--	--	--	--
14	43.6	48.0	52.1	56.8	61.0	66.4	--	--
16	50.1	54.5	58.8	64.2	69.7	76.2	--	--
18	57.6	63.2	68.6	74.0	80.8	88.1	--	--
20	67.6	73.0	78.4	85.0	92.5	100.2	111.6	124.9
22	75.6	81.6	88.1	94.7	103.4	116.5	124.3	133.3
24	92.1	94.9	98.1	104.5	114.2	132.9	140.5	151.2
	4.75	5.00	5.25	5.50	5.75	6.00		
20	133.3	144.0	153.1	164.6	178.3	191.6		
22	144.0	155.6	165.6	176.5	190.9	205.3		
24	162.1	172.8	183.7	194.5	207.1	223.3		

For General Notes on pipe bends, see pages 5 and 6.

PIPE BENDS

Heavy Wall—Over 45° to 90° Inclusive

**Labor Only For Making Pipe Bends With Plain Ends
Carbon Steel Material**

NET MAN HOURS EACH

Nominal Pipe Size	WALL THICKNESS IN INCHES							
	7.50	1.00	1.25	1.50	1.75	2.00	2.25	2.50
4" or Less	6.6	7.5	8.7	10.1	--	--	--	--
5	--	8.6	10.1	11.8	13.5	15.7	--	--
6	--	10.2	11.8	14.2	16.0	19.4	22.6	--
8	--	13.9	14.4	17.0	19.7	23.0	27.2	31.4
10	--	--	17.8	19.5	21.5	24.8	28.7	33.8
12	--	--	--	22.1	23.3	26.9	31.5	36.8
14	--	--	--	27.0	28.2	31.3	36.0	42.5
16	--	--	--	--	34.6	37.3	41.4	48.0
18	--	--	--	--	--	42.5	47.9	56.6
20	--	--	--	--	--	55.8	56.4	65.2
22	--	--	--	--	--	--	--	74.0
24	--	--	--	--	--	--	--	91.1
	2.75	3.00	3.25	3.50	3.75	4.00	4.25	4.50
10	34.9	38.2	--	--	--	--	--	--
12	39.7	43.9	47.2	51.6	--	--	--	--
14	45.8	50.1	54.5	60.0	64.3	69.7	--	--
16	52.3	57.1	63.0	67.9	74.0	80.6	--	--
18	61.2	67.6	73.0	79.0	86.8	93.6	--	--
20	70.8	76.2	82.8	90.4	98.0	106.8	119.5	133.3
22	81.0	88.4	95.8	104.5	114.1	124.2	132.1	144.0
24	96.0	100.2	109.0	119.8	130.7	141.6	149.8	161.0
	4.75	5.00	5.25	5.50	5.75	6.00		
20	144.0	153.1	163.9	175.4	189.1	203.5		
22	154.9	165.6	176.5	187.3	200.3	216.1		
24	172.8	183.7	194.5	208.9	223.3	237.7		

For General Notes on pipe bends, see pages 5 and 6.

For bends 91° through 180° add 75% to the above man hours.

PIPE BENDS

Large O. D. Sizes

Labor Only for Making Pipe Bends with Plain Ends

Carbon Steel Material

MAN HOURS FOR 10′ OF ARC OR PORTION THEREOF

O. D. Pipe Inches	WALL THICKNESS IN INCHES							
	.500 Or Less	.750	1.00	1.25	1.50	1.75	2.00	2.25
26	100.2	101.4	102.0	103.2	104.4	106.2	106.8	115.2
28	114.6	115.8	116.4	118.2	119.4	121.2	123.6	131.4
30	138.6	139.2	140.4	142.2	143.4	145.8	148.2	154.8
32	172.8	174.6	176.4	178.2	180.6	183.6	187.8	194.4
34	216.0	217.8	221.4	223.2	225.6	228.6	231.6	238.2
36	262.8	265.2	266.4	270.0	272.4	277.2	279.0	284.4
	2.50	2.75	3.00	3.25	3.50	3.75	4.00	4.25
26	120.6	126.0	138.8	138.6	144.0	151.2	158.4	167.4
28	135.0	140.4	145.8	151.2	158.4	166.2	176.4	185.4
30	160.8	165.0	171.0	176.4	183.6	189.6	198.1	208.2
32	199.8	205.2	212.4	217.8	225.0	230.4	237.6	244.8
34	229.8	250.2	255.6	261.0	268.2	271.8	277.2	284.4
36	289.8	295.2	300.6	306.0	313.2	319.2	325.8	333.0
	4.50	4.75	5.00	5.25	5.50	5.75	6.00	
26	176.4	189.0	199.8	210.6	223.2	237.6	255.6	
28	195.0	205.2	216.0	228.6	241.2	255.6	271.8	
30	217.8	227.4	239.4	250.2	264.6	281.2	291.6	
32	253.8	262.8	273.0	286.2	297.0	307.8	322.2	
34	291.6	300.6	311.4	320.4	333.0	345.6	360.0	
36	340.2	348.0	357.0	367.2	378.0	388.8	403.2	

For General Notes on pipe bends, see pages 5 and 6.

ATTACHING FLANGES—SCREWED TYPE

Labor—Cutting and Threading Pipe—Making on
Screwed Flanges and Refacing

Carbon Steel Material
For Bends, Headers, Necks and Straight Runs of Pipe

NET MAN HOURS EACH

Pipe Size Inches	125 Lb. Cast Iron and 150 Lb. Steel	250 Lb. Cast Iron and Steel 300 Lb. and Higher
2 or less	1.0	1.2
2-1/2	1.1	1.3
3	1.2	1.4
3-1/2	1.4	1.6
4	1.5	1.7
5	1.6	1.9
6	1.8	2.0
8	2.1	2.4
10	2.6	2.9
12	3.1	3.5
14 OD	3.8	4.3
16 OD	4.6	5.2
18 OD	5.5	6.2
20 OD	6.5	7.4
24 OD	9.3	10.6

Flanges: Man hours are for labor only. The price of the flange must be added in all cases.

Pipe Thickness: Man hours are for any wall thickness of pipe used with listed flange.

Unlisted Sizes: Unlisted sizes take the next higher listing.

ATTACHING FLANGES—SCREWED TYPE

Labor—Cutting and Threading Pipe, Making on Flange,
Manual Seal Welding at Back and Refacing

Carbon Steel Material
Straight Pipe, Bends, Headers and Nozzles

NET MAN HOURS EACH

Size Inches	SERVICE PRESSURE RATING						
	150 Lb.	300 Lb.	400 Lb.	600 Lb.	900 Lb.	1500 Lb.	2500 Lb.
2 or less	1.4	1.5	1.7	1.7	2.2	2.2	2.9
2-½	1.5	1.7	1.9	1.9	2.3	2.3	3.1
3	1.7	1.9	2.0	2.0	2.6	2.6	3.4
3½	1.9	2.1	2.3	2.3	--	--	--
4	2.0	2.3	2.5	2.6	3.0	3.3	3.7
5	2.3	2.6	2.9	3.1	3.4	3.8	4.3
6	2.7	3.0	3.5	3.7	4.1	4.5	5.0
8	3.3	3.7	4.4	4.5	4.9	5.5	6.2
10	4.2	4.7	4.9	5.4	5.9	6.6	7.3
12	4.8	5.4	5.9	6.4	6.9	7.6	8.5
14 OD	5.8	6.5	7.2	8.1	8.9	10.7	--
16 OD	7.2	7.9	8.7	9.6	10.7	11.8	--
18 OD	8.3	9.0	9.7	10.6	11.7	13.1	--
20 OD	9.3	10.1	11.0	12.1	13.3	14.9	--
24 OD	13.0	13.8	14.6	15.6	16.8	18.3	--

Flanges: Man hours are for labor only. The price of the flange must be added in all cases.

Pipe Thickness: Man hours are for any wall thickness of pipe used with listed flange.

Unlisted Sizes: Unlisted sizes take the next higher listing.

ATTACHING FLANGES—SCREWED TYPE

Labor—Cutting and Threading Pipe, Making on Flange,
Manual Seal Welding at Back and Front and Refacing

Carbon Steel Material,
Straight Pipe, Bends, Headers and Nozzles

NET MAN HOURS EACH

Pipe Size Inches	SERVICE PRESSURE RATING						
	150 Lb.	300 Lb.	400 Lb.	600 Lb.	900 Lb.	1500 Lb.	2500 Lb.
2 or less	1.7	1.9	2.1	2.1	2.7	2.7	3.4
2½	1.9	2.1	2.3	2.3	2.9	2.9	3.7
3	2.1	2.3	2.5	2.5	3.2	3.2	4.0
3½	2.3	2.5	2.8	2.8	--	--	--
4	2.5	2.8	3.1	3.3	3.7	4.2	4.6
5	2.9	3.2	3.6	3.9	4.2	4.8	5.3
6	3.4	3.8	4.4	4.6	5.1	5.6	6.1
8	4.1	4.7	5.5	5.6	6.2	6.9	7.6
10	5.3	5.8	6.2	6.8	7.5	8.3	9.3
12	6.0	6.8	7.4	7.9	8.5	9.2	10.0
14 OD	7.2	8.1	8.9	9.9	11.1	12.2	--
16 OD	9.0	9.9	10.9	12.1	13.6	15.0	--
18 OD	10.4	11.3	12.3	13.5	14.7	16.2	--
20 OD	11.6	12.7	13.8	15.2	16.6	18.1	--
24 OD	16.2	17.3	18.2	19.6	21.0	22.5	--

Flanges: Man hours are for labor only. The price of the flange must be added in all cases.

Pipe Thickness: Man hours are for any wall thickness of pipe used with listed flange.

Unlisted Sizes: Unlisted sizes take the next higher listing.

ATTACHING FLANGES—SLIP-ON TYPE

Labor—Slipping on Flange
Manual Welding at Front and Back

Carbon Steel Material

NET MAN HOURS EACH

Size Inches	SERVICE PRESSURE RATING						
	150 Lb.	300 Lb.	400 Lb.	600 Lb.	900 Lb.	1500 Lb.	2500 Lb.
1	0.8	0.9	1.2	1.2	1.4	1.6	1.8
1¼	0.9	1.0	1.2	1.2	1.6	1.8	2.0
1½	0.9	1.1	1.2	1.2	1.6	1.8	2.0
2	1.1	1.2	1.6	1.6	2.1	2.4	2.6
2½	1.3	1.5	2.0	2.0	2.6	2.9	3.2
3	1.6	1.8	2.5	2.5	3.1	3.5	3.9
3½	1.9	2.1	2.9	2.9	--	--	--
4	2.1	2.3	3.0	3.3	4.2	4.7	5.2
5	2.6	2.9	3.9	4.2	5.3	5.9	6.5
6	3.1	3.4	4.5	5.1	6.3	7.1	7.8
8	4.3	4.6	6.2	6.8	8.3	9.3	10.2
10	5.3	5.8	7.6	9.4	10.6	11.9	13.1
12	6.5	7.0	9.3	11.6	13.0	14.6	16.1
14 OD	7.6	8.5	11.0	13.7	15.0	16.8	--
16 OD	8.9	9.6	12.7	15.6	17.0	19.0	--
18 OD	10.3	11.4	14.8	17.9	20.1	22.5	--
20 OD	12.4	13.6	17.9	20.1	23.3	26.1	--
24 OD	15.5	17.0	21.7	26.4	29.5	33.0	--
26 OD	-	-	23.5	28.6	32.0	-	-
30 OD	-	-	27.1	33.0	36.9	-	-
34 OD	-	-	30.7	37.4	41.8	-	-
36 OD	-	-	32.5	39.6	44.2	-	-
42 OD	-	-	37.9	46.2	-	-	-

Flanges: Man hours are for labor attaching the flange perpendicular to the centerline of a section of straight pipe or to a straight section of pipe on the end of a bend.

Pipe Thickness: Man hours are for any wall thickness of pipe used with listed flanges.

Weld Fittings or Bends: For attaching flanges to weld fittings or bends with no straight tangents or on a straight section of pipe, but other than perpendicular to the centerline, add 25% to the above man hours.

Refacing: For flanges requiring refacing after welding increase above man hours 50%.

ATTACHING FLANGES—SLIP-ON TYPE

Labor—Slipping on Flange and
Machine Welding at Front and Back

Carbon Steel Material

NET MAN HOURS EACH

Size Ins.	SERVICE PRESSURE RATING						
	150 Lb.	300 Lb.	400 Lb.	600 Lb.	900 Lb.	1500 Lb.	2500 Lb.
2	.66	.72	.96	.96	1.26	1.44	1.56
2-1/2	.78	.90	1.20	1.20	1.56	1.74	1.92
3	.96	1.08	1.50	1.50	1.86	2.10	2.34
4	1.26	1.38	1.80	1.98	2.52	2.82	3.12
6	1.55	1.70	2.25	2.55	3.15	3.55	3.90
8	1.72	1.84	2.48	2.72	3.32	3.72	4.10
10	2.12	2.32	3.04	3.76	4.24	4.76	5.24
12	2.18	2.45	3.26	4.06	4.55	5.11	5.64
14 OD	2.28	2.55	3.30	4.11	4.80	5.88	--
16 OD	2.67	2.88	3.81	4.68	5.10	6.65	--
18 OD	3.09	3.42	4.40	5.37	6.03	6.75	--
20 OD	3.72	4.08	5.37	6.03	6.99	7.83	--
24 OD	4.65	5.10	6.51	7.92	8.85	9.90	--
26 OD	-	-	7.05	8.58	9.59	-	-
30 OD	-	-	8.14	9.90	11.06	-	-
34 OD	-	-	9.22	11.22	12.54	-	-
36 OD	-	-	9.76	11.88	13.28	-	-
42 OD	-	-	11.39	13.86	-	-	-

Man hours include slipping-on carbon steel flange, tack welding and machine submerged arc. Welding both front and back. For sizes 2″ through 4″ time is included for manual welding on front.

Man hours are for any wall thickness of pipe used with listed flange.

Above man hours should be used in lieu of manual welding slip-on flanged joints on all shop machine welded slip-on flanges which can be rotated.

Unlisted sizes take the next highest listing.

For additional information see notes at bottom of preceding page.

ATTACHING FLANGES—WELD NECK TYPE

Labor—Aligning Flange and Butt Welding

Carbon Steel Material
NET MAN HOURS EACH

Size Ins.	SERVICE PRESSURE RATING						
	150 Lb.	300 Lb.	400 Lb.	600 Lb.	900 Lb.	1500 Lb.	2500 Lb.
2	1.05	1.20	1.20	1.50	1.50	1.60	1.70
2½	1.40	1.60	1.60	1.80	1.80	1.90	2.20
3	1.70	1.90	1.90	2.20	2.20	2.30	2.40
4	2.10	2.30	2.30	2.60	2.60	2.90	3.00
6	2.90	3.20	3.20	3.50	3.50	3.90	4.00
8	3.60	3.90	3.90	4.30	4.30	4.90	5.10
10	4.50	4.80	4.80	5.30	5.30	6.10	6.20
12	5.25	5.70	5.70	6.30	6.30	7.40	8.00
14 OD	6.10	6.60	6.60	7.40	7.40	8.90	—
16 OD	7.10	7.70	7.70	8.60	8.60	11.30	—
18 OD	8.30	8.90	8.90	10.10	10.10	12.00	—
20 OD	9.10	9.90	9.90	11.20	11.20	14.30	—
24 OD	10.20	11.10	11.10	12.70	12.70	16.30	—
26 OD	—	—	12.03	13.76	13.76	—	—
30 OD	—	—	13.88	15.88	15.88	—	—
34 OD	—	—	15.73	18.00	18.00	—	—
36 OD	—	—	16.65	19.06	19.06	—	—
42 OD	—	—	19.43	22.23	—	—	—

Man hours include aligning and tack welding carbon steel weld neck flange and machine submerged arc butt welding to pipe.

Man hours are for any wall thickness of pipe used with listed flanges.

Unlisted sizes take the next highest listing.

ATTACHING ORIFICE FLANGES— SLIP-ON AND THREADED TYPES

Carbon Steel Material

MAN HOURS PER PAIR

Size Ins.	SERVICE PRESSURE RATING			
	Slip-On Type		Threaded Types	
	300 Lb.	300 Lb.	400–600 Lb.	900–1500 Lb.
3	5.3	4.7	–	8.3
4	6.5	6.5	7.4	9.5
6	9.0	8.0	9.3	12.0
8	12.2	9.9	12.9	14.8
10	15.5	12.8	15.8	18.2
12	18.3	16.1	18.9	21.6
14	22.0	18.5	–	–
16	25.1	21.2	–	–
18	28.7	24.5	–	–
20	33.9	27.8	–	–
24	41.0	35.3	–	–
26	53.4	–	–	–
30	60.5	–	–	–
34	69.0	–	–	–
36	74.1	–	–	–
42	80.1	–	–	–

Slip-On Types: Man hours include slipping on, welding, placement of paddle-type plates, and bolting of pair of orifice flanges.

Threaded Types: Man hours include screwing on, placement of paddle-type plates, and bolting up of pair of orifice-type flanges.

All man hours exclude cutting, beveling, or threading of pipe. See respective tables for these man hours.

ATTACHING ORIFICE FLANGES—WELD NECK TYPE

Carbon Steel Material

MAN HOURS PER PAIR

Size Ins.	SERVICE PRESSURE RATING				
	300 Lb.	400 Lb.	600 Lb.	900 Lb.	1500 Lb.
3	5.6	6.0	7.7	7.8	8.3
4	7.8	8.2	9.4	9.5	10.7
6	9.8	10.4	12.2	12.4	13.8
8	13.1	13.4	15.6	15.9	18.2
10	14.7	15.0	19.4	20.2	21.6
12	16.8	18.8	20.3	21.5	24.3
14	19.2	20.5	22.8	24.1	27.8
16	21.2	22.0	27.6	28.2	31.2
18	26.1	26.9	29.7	31.1	35.9
20	28.7	29.4	33.3	34.1	41.6
24	37.1	37.8	43.3	44.2	—
26	—	41.1	44.9	45.6	—
30	—	47.5	50.9	52.6	—
34	—	57.2	60.7	62.6	—
36	—	60.0	67.4	68.7	—
42	—	87.6	91.7	—	—

Man hours include setting, aligning, welding, placement of paddle-type plates, and bolting up of pair of orifice flanges.

Man hours exclude cutting and beveling of pipe. See respective tables for these man hours.

GENERAL WELDING NOTES

Backing Rings: When backing rings are used, add 25% to the welding man hours to cover extra problems in fit-up. In addition the following percentages should be added if applicable:

    1) When backing rings are tack welded in on one side, add 10% to the man hours of a standard thickness butt weld.

    2) When backing rings are completely welded in on one side, add 30% to the man hours of a standard thickness butt weld.

    3) Preheating and stress relieving, when required, should be charged at full butt weld preheating and stress relieving man hours for the size and thickness in which the backing ring is installed.

Nozzle Welds: Following percentage increases should be allowed for the following conditions:

    1) When nozzle welds are to be located off-center of the run(except tangential) increase man hours shown for nozzle welds, 50%.

    2) Add 80% to nozzle welds for tangential nozzle welds.

    3) When nozzle welds are to be located on a fitting increase nozzle weld man hours 50%.

Long Neck Nozzle Welds: The welding-on of long neck nozzles should be charged at the schedule 160 reinforced nozzle weld man hours.

Shaped Nozzles, Nozzle Weld Fit-Ups and Dummy Nozzle Welds: These should be charged at a percentage of the completed nozzle weld man hours as follows:

    1) Shaped Branch 50%
    2) Shaped Hole in Header 50%
    3) Fit-up of Both Branch or Header (whether tack-welded or not) 60%
    4) Dummy Nozzle Weld (no holes in header) 70%

Sloping Lines: Add 100% to all welding man hours for this condition.

Consumable Inserts: When consumable inserts are used, add the following percentages to the welding man hours to cover extra problems in fit-up:

    1) Through 1/2" wall 40%
    2) Over 1/2" through 1" wall 30%
    3) Over 1" through 2" wall 20%
    4) Over 2" through 3" wall 15%
    5) Over 3" wall 10%

SPECIAL FITTING AND PREPARATION FOR INERT GAS SHIELDED ROOT PASS WELDING

Butt Welds

NET MAN HOURS EACH

CARBON STEEL, CHROME ALLOY AND STAINLESS STEEL

Pipe Size Inches	All Thicknesses	Pipe Size Inches	All Thicknesses
2" or less	0.45	26 O.D.	5.37
2-1/2	0.64	28 O.D.	6.12
3	0.75	30 O.D.	6.48
4	0.94	32 O.D.	6.83
5	1.14	34 O.D.	6.88
6	1.36	36 O.D.	7.06
8	1.67	38 O.D.	7.39
10	1.92	40 O.D.	8.23
12	2.06	42 O.D.	9.06
14 O.D.	2.66	44 O.D.	9.87
16 O.D.	3.19	46 O.D.	10.93
18 O.D.	3.90	48 O.D.	11.99
20 O.D.	4.22	54 O.D.	13.15
24 O.D.	4.68	60 O.D.	14.42

Man hours shown will apply either with or without an internal nitrogen purge.

For internal argon purge, increase above man hours 20 percent.

Man hours do not include the use of an oxygen analyzer.

Man hours do not include the installation of consumable inserts or end preparation for consumable inserts.

For preparation of nozzle welds, 'olet welds, coupling welds, and mitre butt welds for inert gas shielded root pass add 100 percent to the above man hours.

If the purge is to be held longer than the first two passes, increase the above man hours 50 percent for each additional pass for which the purge is held.

MACHINE BUTT WELDS

Submerged Arc Butt Welds
Carbon Steel Material
NET MAN HOURS EACH

| Size | SCHEDULE NUMBER | | | |
Ins.	40	60	80	160
2	.40	--	.45	.70
2-1/2	.50	--	.55	.80
3	.55	--	.60	.90
4	.65	--	.80	1.30
6	.90	--	1.05	2.15
8	.99	1.0	1.12	2.92
10	1.08	1.36	1.72	4.44
12	1.19	1.54	1.96	5.32
14 O.D.	1.26	1.74	2.43	--
16 O.D.	1.68	2.13	3.15	--
18 O.D.	2.19	2.85	4.17	--
20 O.D.	2.37	3.51	4.95	--
24 O.D.	3.36	5.10	7.89	--

Man hours include cutting, beveling, fitting, tack welding, manual single pass or backing ring, machine set-up and submerged welding.

Above man hours should be used in lieu of manual butt weld man hours on all shop machine welds which can be rotated.

Above man hours do not include preheating, grinding or stress relieving. See respective tables for these charges.

This procedure not applicable to alloy pipe.

All sizes of butt welds shown below the ruled lines are ¾" or greater in wall thickness and must be stress relieved.

MANUAL BUTT WELDS

Labor for Welding Only
Carbon Steel Materials
NET MAN HOURS EACH

Size Ins.	Standard Pipe & OD Sizes 3/8" Thick	Extra Heavy Pipe & OD Sizes 1/2" Thick	SCHEDULE NUMBERS								
			20	30	40	60	80	100	120	140	160
1	0.6	0.7	--	--	0.6	--	0.7	--	--	--	0.9
1-1/4	0.7	0.7	--	--	0.7	--	0.7	--	--	--	1.0
1-1/2	0.7	0.8	--	--	0.7	--	0.8	--	--	--	1.1
2	0.8	0.9	--	--	0.8	--	0.9	--	--	--	1.4
2-1/2	1.0	1.1	--	--	1.0	--	1.1	--	--	--	1.6
3	1.1	1.2	--	--	1.1	--	1.2	--	--	--	1.8
3-1/2	1.2	1.4	--	--	1.2	--	1.4	--	--	--	--
4	1.3	1.6	--	--	1.3	--	1.6	--	2.4	--	2.6
5	1.5	1.9	--	--	1.5	--	1.9	--	2.5	--	3.3
6	1.8	2.1	--	--	1.8	--	2.1	--	3.2	--	4.3
8	2.2	2.8	2.2	2.2	2.2	2.5	2.8	3.8	5.1	6.4	7.3
10	2.7	3.4	2.7	2.7	2.7	3.4	4.3	5.8	8.0	9.6	11.1
12	3.1	4.0	3.1	3.1	3.4	4.4	5.6	8.4	10.4	13.0	15.2
14 OD	3.6	4.8	3.6	3.6	4.2	5.8	8.1	11.2	13.7	16.3	19.2
16 OD	4.2	5.6	4.2	4.2	5.6	7.1	10.5	14.3	17.6	21.1	23.5
18 OD	4.9	6.5	4.9	5.7	7.3	9.5	13.9	18.5	21.7	25.4	28.6
20 OD	5.3	7.1	5.3	7.1	7.9	11.7	16.5	22.1	27.0	31.3	34.6
24 OD	5.9	8.5	5.9	10.5	11.2	17.0	26.3	30.4	36.9	41.8	50.2

Pipe Thickness: Wall thickness of the pipe determines the man hours that will apply. For butt welds of double extra strong materials, use Schedule 160 listings.

Mitre Welds: Add 50% to butt weld man hours.

Cutting and Beveling Pipe: Man hours do not include cutting and beveling of pipe. See respective tables for these charges.

Preheating: If specified or required by Codes, add for this operation. See man hours for preheating.

Stress Relieving: Stress relieving of welds in carbon steel material is required by the A.S.A. Code of Pressure piping where the wall thickness is 3/4" or greater.

All sizes of butt welds shown below the ruled lines are 3/4" or greater in wall thickness and must be stress relieved.

Where stress relieving is required, an extra charge should be made. See man hours for stress relieving.

Unlisted Sizes: Unlisted sizes take the next higher listing.

General Notes: For additional notes on welding see page 18.

MANUAL HEAVY WALL BUTT WELDS

Labor for Welding Only

Carbon Steel Material

NET MAN HOURS EACH

Nominal Pipe Size	WALL THICKNESS IN INCHES							
	.750	1.00	1.25	1.50	1.75	2.00	2.25	2.50
3	2.3	3.1	--	--	--	--	--	--
4	2.8	3.5	4.8	5.8	--	--	--	--
5	--	4.0	5.7	6.8	8.5	10.5	--	--
6	--	5.4	7.2	8.8	11.3	13.2	15.4	--
8	--	7.4	8.6	11.1	14.0	16.3	19.2	23.2
10	--	--	11.4	13.7	17.0	19.7	23.1	27.2
12	--	--	--	16.6	19.7	23.2	27.5	31.8
14	--	--	--	19.9	22.5	26.4	30.9	36.5
16	--	--	--	--	25.3	30.2	35.2	42.1
18	--	--	--	--	--	33.7	39.3	46.4
20	--	--	--	--	--	39.3	46.4	56.1
22	--	--	--	--	--	--	--	61.3
24	--	--	--	--	--	--	--	66.7

	2.75	3.00	3.25	3.50	3.75	4.00	4.25	4.50
10	31.1	35.7	--	--	--	--	--	--
12	36.3	41.6	46.9	53.5	--	--	--	--
14	41.4	47.0	53.3	60.4	68.8	77.2	--	--
16	47.7	54.8	61.8	70.2	80.8	91.2	--	--
18	53.3	61.8	70.2	80.5	91.9	105.2	--	--
20	63.9	71.6	82.1	92.6	105.2	119.3	135.3	146.7
22	69.8	78.6	89.4	101.1	114.9	130.6	148.1	163.9
24	75.8	83.9	96.9	109.5	124.9	141.9	160.8	177.4

	4.75	5.00	5.25	5.50	5.75	6.00		
20	160.8	172.3	183.8	191.4	213.1	227.2		
22	173.6	186.3	203.9	214.1	234.6	250.1		
24	189.4	203.9	222.8	233.5	252.7	270.3		

For General Notes on welding, see pages 20 and 23.

MANUAL LARGE O.D. BUTT WELDS

Labor for Welding Only

Carbon Steel Material

NET MAN HOURS EACH

O.D. PIPE INCHES	WALL THICKNESS IN INCHES							
	.375	.500	.750	1.00	1.25	1.50	1.75	2.00
26	7.1	9.7	12.8	17.1	22.6	29.2	36.8	44.5
28	8.5	11.1	13.9	18.8	24.8	31.6	39.3	47.2
30	10.6	12.9	16.0	20.4	26.9	33.7	42.0	49.9
32	13.1	15.2	18.2	22.6	29.6	36.4	44.7	52.6
34	16.4	18.2	20.7	25.0	33.3	39.1	47.7	55.4
36	19.5	20.9	23.6	28.1	38.3	44.1	52.8	60.8
38	22.9	24.5	27.1	31.4	44.1	49.8	58.1	66.2
40	26.8	29.0	31.2	35.2	50.6	56.2	63.9	72.2
42	31.3	34.2	36.0	39.5	58.3	63.6	70.2	78.7
44	36.3	39.5	42.3	48.3	63.5	70.5	76.5	85.8
46	40.9	45.0	49.4	57.5	70.1	77.5	83.3	92.9
48	46.2	50.8	57.7	67.0	77.0	84.7	90.5	100.1
54	52.0	57.3	67.4	78.1	84.5	92.6	98.3	107.9
60	58.5	64.6	78.7	91.0	92.8	101.2	106.8	116.3
	2.75	2.50	2.75	3.00	3.25	3.50	3.75	4.00
26	52.3	72.0	81.6	93.2	104.7	117.4	135.3	152.9
28	55.1	77.1	88.5	99.6	117.1	127.4	148.1	165.4
30	57.9	84.3	95.7	107.2	122.2	136.5	157.0	177.4
32	60.5	88.1	100.8	112.6	129.7	144.2	166.9	188.9
34	65.2	93.2	107.2	120.5	136.8	152.9	177.4	201.6
36	70.3	99.6	114.3	127.7	145.5	162.9	188.6	211.9
	4.25	4.50	4.75	5.00	5.25	5.50	5.75	6.00
26	172.3	191.4	206.7	221.8	234.6	252.7	274.3	293.0
28	186.3	207.5	221.5	237.4	253.9	272.9	292.8	311.4
30	199.1	219.0	238.6	255.2	273.1	291.0	314.5	339.0
32	211.9	234.8	252.7	270.6	291.0	310.0	334.4	359.9
34	227.2	252.7	270.6	288.9	310.9	331.8	357.3	382.8
36	239.4	265.4	285.9	305.8	328.0	349.7	377.7	405.3

For General Notes on welding, see pages 20 and 23.

90° WELDED NOZZLES

Labor For Cutting And Welding
Carbon Steel Material

NET MAN HOURS EACH

Size Ins.	Standard Pipe & OD Sizes 3/8" Thick	Extra Heavy Pipe & OD Sizes 1/2" Thick	SCHEDULE NUMBERS								
			20	30	40	60	80	100	120	140	160
1	1.8	1.9	--	--	1.8	--	1.9	--	--	--	2.7
1-1/4	1.9	2.1	--	--	1.9	--	2.1	--	--	--	3.1
1-1/2	2.1	2.3	--	--	2.1	--	2.3	--	--	--	3.5
2	2.2	2.7	--	--	2.2	--	2.7	--	--	--	4.6
2-1/2	2.4	3.3	--	--	2.4	--	3.3	--	--	--	5.1
3	2.8	3.8	--	--	2.8	--	3.8	--	--	--	5.7
3-1/2	3.2	4.3	--	--	3.2	--	4.3	--	--	--	--
4	3.5	4.9	--	--	3.5	--	4.9	--	6.1	--	7.5
5	4.4	6.0	--	--	4.4	--	6.0	--	7.5	--	9.3
6	4.7	6.5	--	--	4.7	--	6.5	--	9.5	--	12.1
8	5.3	7.5	5.3	5.3	5.3	7.0	7.5	10.2	12.9	15.8	18.2
10	6.0	8.7	6.0	6.0	6.0	8.7	10.7	13.9	17.9	23.1	27.8
12	6.9	10.0	6.9	6.9	8.4	11.1	14.4	19.9	24.3	29.4	33.2
14 OD	7.9	11.5	7.9	7.9	9.8	13.6	19.2	24.5	29.3	33.0	40.6
16 OD	9.0	12.9	9.0	9.0	12.9	17.1	22.7	30.9	35.1	38.8	46.9
18 OD	9.8	13.8	9.8	12.8	16.2	21.6	25.5	37.4	41.5	45.0	58.7
20 OD	11.0	15.5	11.0	15.5	18.9	27.6	30.0	43.3	47.3	51.6	65.9
24 OD	12.0	16.8	12.0	18.0	23.5	35.3	39.0	55.0	59.0	66.0	77.1

All Nozzles other than 90° should be charged at the man hours shown for 45° nozzles.

Pipe Thickness: Wall thickness of the pipe used for the nozzle determines the man hours that will apply. For nozzle of double extra strong pipe thickness, use Schedule 160 man hours.

Reinforcement: Man hours given above are for plain welded nozzles only. For use of Gusset plates, etc., as stiffeners not for reinforcement, add 25% to the net man hours shown above. If reinforcement is required and produced by building up the nozzle weld, or by the use of reinforcing rings or saddles as specified, use man hours for 90° reinforced nozzles.

Preheating: If specified or required by Codes, add for this operation. See man hours for preheating. The size and wall thickness of header (not the size of the nozzle) determines the preheating man hours.

Stress Relieving: Stress relieving of welds in carbon steel material is required by the A. S. A. Code for Pressure Piping, where the wall thickness is 3/4" or greater. The size and wall thickness of the header determines the man hours to be used for stress relieving.

All pipe sizes shown below the ruled line are 3/4" or greater in wall thickness and must be stress relieved. Where stress relieving is required an extra charge should be made. See man hours for stress relieving.

For General Notes on welding, see pages 20 and 23.

90° WELDED NOZZLES—REINFORCED

Labor For Cutting and Welding

Carbon Steel Material

NET MAN HOURS EACH

Size Ins.	Standard Pipe & OD Sizes 3/8" Thick	Extra Heavy Pipe & OD Sizes 1/2" Thick	SCHEDULE NUMBERS								
			20	30	40	60	80	100	120	140	160
1-1/2	4.4	4.7	--	--	4.4	--	4.7	--	--	--	6.7
2	4.6	5.1	--	--	4.6	--	5.1	--	--	--	8.6
2-1/2	5.0	6.1	--	--	5.0	--	6.1	--	--	--	9.4
3	5.8	6.9	--	--	5.8	--	6.9	--	--	--	10.4
3-1/2	6.5	7.6	--	--	6.5	--	7.6	--	--	--	--
4	7.0	8.7	--	--	7.0	--	8.7	--	10.9	--	13.3
5	8.5	10.3	--	--	8.5	--	10.3	--	13.1	--	16.0
6	8.9	11.1	--	--	8.9	--	11.1	--	16.1	--	20.1
8	9.9	12.3	9.9	9.9	9.9	11.3	12.3	16.8	20.8	24.8	27.8
10	10.8	13.8	10.8	10.8	10.8	13.8	17.1	22.1	26.3	29.5	33.1
12	12.0	15.5	12.0	12.0	13.0	17.2	22.3	30.1	33.5	37.8	40.4
14 OD	13.6	17.5	13.6	13.6	15.1	20.6	29.2	37.3	39.7	44.6	51.3
16 OD	15.2	19.7	15.2	15.2	19.3	25.5	34.0	46.4	48.7	51.4	59.3
18 OD	16.2	20.2	16.2	18.7	23.7	31.7	38.1	50.4	52.4	55.4	74.2
20 OD	17.9	22.5	17.9	22.4	27.4	40.0	47.0	54.0	62.5	71.8	83.3
24 OD	18.9	23.5	18.9	25.7	30.9	44.5	50.4	61.3	72.6	84.0	97.4

All Nozzles other than 90° should be charged at the man hours shown for 45° nozzles.

Pipe Thickness: Wall thickness of the pipe used for the nozzle determines the man hours that will apply. For nozzles of double extra strong pipe thickness, use Schedule 160 man hours.

Reinforcement: Man hours are for labor only. The price of the nozzle and reinforcing materials must be added.

Preheating: If specified or required by Codes, add for this operation. See man hours for preheating. The size and wall thickness of header (not the size of the nozzle) determines the preheating man hours.

Stress Relieving: Stress relieving of welds in carbon steel material is required by the A.S.A. Code for Pressure Piping, where the wall thickness is 3/4" or greater. The size and wall thickness of the header determines the man hours to be used for stress relieving.

All pipe sizes shown below the ruled line are 3/4" or greater in wall thickness and must be stress relieved. Where stress relieving is required an extra charge should be made. See man hours for stress relieving.

For General Notes on welding, see pages 20 and 23.

LARGE O.D. 90° NOZZLE WELDS

Labor For Cutting And Welding

Carbon Steel Material

NET MAN HOURS EACH

O. D. Pipe Inches	NON REINFORCED 90° NOZZLE WELDS								
	.375	.500	.750	1.00	1.25	1.50	1.75	2.00	2.25
26	21.0	24.6	27.0	38.5	45.2	58.0	66.9	75.4	88.6
28	23.8	28.1	29.4	42.4	49.7	63.3	71.3	79.9	93.3
30	28.1	32.4	33.9	45.9	53.8	67.3	76.1	84.5	98.1
32	32.8	36.7	38.9	50.7	59.1	72.9	81.0	88.9	102.3
34	39.5	40.8	43.8	56.3	66.6	78.3	86.5	93.7	110.3
36	45.6	46.7	50.1	63.3	76.5	88.0	95.8	102.8	119.0
38	51.5	52.7	57.6	71.5	88.1	99.5	106.4	112.1	128.6
40	58.2	59.6	66.3	80.8	101.3	112.4	118.1	122.2	138.9
42	65.8	67.3	76.2	91.3	116.5	127.0	131.1	133.1	150.0
48	72.0	76.8	86.9	104.2	133.0	145.0	149.8	152.2	171.4
54	81.0	86.4	97.7	117.2	149.6	163.1	168.5	171.2	192.8
60	90.0	96.0	108.6	130.2	166.2	181.2	187.2	190.2	214.2

	REINFORCED 90° NOZZLE WELDS								
	.375	.500	.750	1.00	1.25	1.50	1.75	2.00	2.25
26	29.1	34.2	37.4	45.4	53.0	68.5	86.5	104.5	123.0
28	32.9	39.0	40.8	49.6	58.0	74.2	92.3	111.0	129.4
30	38.9	44.9	47.0	53.9	63.2	79.2	98.7	117.3	136.1
32	45.4	50.7	53.5	59.8	69.5	85.5	105.0	123.6	142.1
34	54.9	56.7	60.8	66.1	78.2	92.0	112.3	130.0	153.3
36	63.2	64.8	69.3	74.2	89.9	103.5	123.9	142.7	165.1
38	68.7	73.9	79.1	83.9	103.4	117.0	136.4	155.6	178.3
40	80.8	84.2	90.2	94.8	119.0	132.2	150.0	169.6	192.6
42	91.2	96.0	102.8	107.1	136.8	149.4	165.1	184.9	208.1
48	104.2	109.9	117.6	122.4	156.5	170.9	188.6	211.2	237.6
54	117.2	123.7	132.3	137.7	176.0	192.2	212.2	237.6	267.3
60	130.2	137.4	147.0	153.0	195.6	213.6	235.8	264.0	297.0

For General Notes on welding, see pages 20, 23, 26, and 27.

45° WELDED NOZZLES

Labor For Cutting And Welding

Carbon Steel Material

NET MAN HOURS EACH

Size Ins.	Standard Pipe & OD Sizes 3/8" Thick	Extra Heavy Pipe & OD Sizes 1/2" Thick	SCHEDULE NUMBERS								
			20	30	40	60	80	100	120	140	160
1	2.4	2.5	--	--	2.4	--	2.5	--	--	--	3.6
1-1/4	2.5	2.8	--	--	2.5	--	2.8	--	--	--	4.1
1-1/2	2.8	3.1	--	--	2.8	--	3.1	--	--	--	4.7
2	2.9	3.6	--	--	2.9	--	3.6	--	--	--	6.2
2-1/2	3.3	4.4	--	--	3.3	--	4.4	--	--	--	6.9
3	3.8	5.1	--	--	3.8	--	5.1	--	--	--	7.7
3-1/2	4.3	5.6	--	--	4.3	--	5.6	--	--	--	--
4	4.8	6.6	--	--	4.8	--	6.6	--	8.2	--	9.9
5	5.8	7.9	--	--	5.8	--	7.9	--	10.2	--	12.5
6	6.2	8.7	--	--	6.2	--	8.7	--	12.6	--	16.0
8	7.3	9.9	7.3	7.3	7.3	9.4	9.9	13.5	17.0	21.3	24.2
10	8.1	11.6	8.1	8.1	8.1	11.6	14.1	18.5	23.8	30.7	37.0
12	9.2	13.3	9.2	9.2	11.0	14.8	19.4	20.5	32.7	38.7	44.2
14 OD	10.5	15.4	10.5	10.5	13.0	18.1	25.5	32.5	38.9	43.7	54.4
16 OD	12.0	17.1	12.0	12.0	17.1	22.7	29.9	41.3	46.4	51.8	62.8
18 OD	13.1	18.3	13.1	17.0	21.6	29.1	34.4	50.1	55.6	60.1	78.6
20 OD	14.5	20.1	14.5	20.1	25.1	36.7	39.7	57.9	63.4	70.0	88.3
24 OD	15.9	22.5	15.9	28.2	30.1	46.8	47.5	73.6	79.1	89.6	103.3

Pipe Thickness: Wall thickness of the pipe used for the nozzle determines the man hours that will apply. For nozzles of double extra strong pipe thickness, use Schedule 160 man hours.

Reinforcement: Man hours given above are for plain nozzles only. For use of Gusset plates, etc., as stiffeners, not for reinforcement, add 25% to the net man hours shown above. If reinforcement is required and produced by building up the nozzle weld, or by the use of reinforcing rings or saddles as specified use man hours for 45° reinforced nozzles.

Preheating: If specified or required by Codes, add for this operation. See man hours for preheating. The size and wall thickness of header (not the size of the nozzle) determines the preheating man hours.

Stress Relieving: Stress relieving of welds in carbon steel material is required by the A. S. A. Code for Pressure Piping, where the wall thickness is 3/4" or greater. The size and wall thickness of the header determines the man hours to be used for stress relieving.

All pipe sizes shown below the ruled line are 3/4" or greater in wall thickness and must be stress relieved. Where stress relieving is required an extra charge should be made. See man hours for stress relieving.

For General Notes on welding, see pages 20 and 23.

45° WELDED NOZZLES—REINFORCED

Labor For Cutting And Welding

Carbon Steel Material

NET MAN HOURS EACH

Size Ins.	Standard Pipe & OD Sizes 3/8" Thick	Extra Heavy Pipe & OD Sizes 1/2" Thick	SCHEDULE NUMBERS								
			20	30	40	60	80	100	120	140	160
1-1/2	5.9	6.2	--	--	5.9	--	6.2	--	--	--	8.9
2	6.0	6.7	--	--	6.0	--	6.7	--	--	--	11.6
2-1/2	6.8	8.1	--	--	6.8	--	8.1	--	--	--	12.7
3	7.6	9.1	--	--	7.6	--	9.1	--	--	--	14.0
3-1/2	8.7	10.1	--	--	8.7	--	10.1	--	--	--	--
4	9.7	11.7	--	--	9.7	--	11.7	--	14.5	--	17.6
5	11.4	13.7	--	--	11.4	--	13.7	--	17.2	--	21.7
6	12.0	14.9	--	--	12.0	--	14.9	--	21.3	--	27.3
8	13.4	18.3	13.4	13.4	13.4	15.3	18.3	22.4	27.8	33.7	37.7
10	14.7	18.4	14.7	14.7	14.7	18.4	22.0	27.2	37.7	41.4	45.1
12 OD	16.3	20.6	16.3	16.3	17.2	23.0	30.0	38.7	43.8	51.5	60.1
14 OD	17.8	23.4	17.8	17.8	19.8	27.6	38.9	44.6	52.7	59.7	68.8
16 OD	20.1	25.6	20.1	20.1	25.6	34.0	44.8	55.7	70.0	71.8	79.5
18 OD	21.6	26.8	21.6	24.9	31.7	42.5	50.4	67.6	70.3	74.1	99.5
20 OD	23.9	30.2	23.9	30.2	36.9	53.3	60.4	72.5	83.9	96.0	111.8
24 OD	25.0	32.5	25.0	32.8	43.3	60.5	65.1	82.3	97.4	112.3	130.6

Pipe Thickness: Wall thickness of the pipe used for the nozzle determines the man hours that will apply. For nozzles of double extra strong pipe thickness, use Schedule 160 man hours.

Reinforcement: Man hours are for labor only. The price of the nozzle and reinforcing materials must be added.

Preheating: If specified or required by Codes, add for this operation. See man hours for preheating. The size and wall thickness of header (not the size of the nozzle) determines the preheating man hours.

Stress Relieving: Stress relieving of welds in carbon steel material is required by the A.S.A. Code for Pressure Piping, where the wall thickness is 3/4" or greater. The size and wall thickness of the header determines the man hours to be used for stress relieving.

All pipe sizes shown below the ruled line are 3/4" or greater in wall thickness and must be stress relieved. Where stress relieving is required an extra charge should be made. See man hours for stress relieving.

For General Notes on welding, see pages 20 and 23.

LARGE O.D. 45° NOZZLE WELDS

Labor for Cutting and Welding

Carbon Steel Material

NET MAN HOURS EACH

O.D. Pipe Inches	NON-REINFORCED 45° NOZZLE WELDS								
	.375	.500	.750	1.00	1.25	1.50	1.75	2.00	2.25
26	28.1	32.8	36.1	51.5	60.2	77.8	89.1	100.5	118.2
28	31.5	37.6	39.1	56.4	65.8	84.2	95.1	106.5	124.4
30	37.4	43.2	45.1	61.2	71.7	90.0	101.6	112.7	130.7
32	43.6	48.8	51.4	67.9	78.8	97.2	108.0	118.8	136.5
34	52.7	54.6	58.3	75.0	88.8	104.3	115.5	124.9	147.3
36	60.7	62.2	66.7	84.2	102.0	117.4	127.5	137.2	158.5
38	68.6	71.0	76.8	95.2	117.4	132.7	140.4	149.6	171.4
40	77.5	80.9	88.3	107.6	135.1	149.9	154.4	163.0	185.0
42	87.6	92.2	101.6	121.5	155.3	169.4	176.1	190.4	199.8
48	100.3	105.6	116.2	138.7	177.6	192.0	201.1	217.4	228.0
54	112.9	118.8	130.7	156.1	199.8	216.0	226.3	244.6	256.5
60	125.4	132.0	145.2	173.4	222.0	240.0	251.4	271.8	285.0

REINFORCED 45° NOZZLE WELDS

	.375	.500	.750	1.00	1.25	1.50	1.75	2.00	2.25
26	38.9	45.5	49.9	63.6	70.6	91.2	115.3	139.3	163.8
28	43.7	52.0	54.3	66.1	77.4	99.0	123.1	147.9	172.7
30	51.8	59.9	62.7	71.9	84.2	105.8	131.5	156.3	181.4
32	60.6	67.6	71.4	79.7	92.7	114.0	140.0	164.8	189.4
34	73.1	75.7	81.0	88.1	104.3	122.6	149.7	173.4	204.3
36	84.1	86.3	92.5	99.0	119.9	137.9	165.2	190.2	220.2
38	95.0	98.5	105.5	111.9	137.9	155.8	181.9	207.4	237.9
40	107.4	112.2	120.3	126.4	158.6	176.1	200.0	226.0	256.9
42	121.4	127.9	137.2	142.9	182.4	199.0	220.0	246.4	277.5
48	138.7	146.4	157.0	163.2	208.3	227.5	251.5	281.8	317.3
54	156.1	164.7	176.6	183.6	234.4	256.0	283.0	317.0	356.9
60	173.4	183.0	196.2	204.0	260.4	284.4	314.4	352.2	396.6

For General Notes on welding, see pages 20, 23, 29, and 30.

CONCENTRIC SWEDGED ENDS

Labor for Welding

Carbon Steel Material

NET MAN HOURS EACH

Size Ins.	Standard Pipe & OD Sizes 3/8" Thick	Extra Heavy Pipe & OD Sizes 1/2" Thick	SCHEDULE NUMBERS								
			20	30	40	60	80	100	120	140	160
2	1.4	1.8	--	--	1.4	--	1.8	--	--	--	2.8
2-1/2	1.6	2.1	--	--	1.6	--	2.1	--	--	--	3.3
3	1.7	2.4	--	--	1.7	--	2.4	--	--	--	3.9
3-1/2	2.0	2.8	--	--	2.0	--	2.8	--	--	--	--
4	2.3	3.3	--	--	2.3	--	3.3	--	4.7	--	5.6
5	3.0	4.2	--	--	3.0	--	4.2	--	6.6	--	7.8
6	3.6	5.4	--	--	3.6	--	5.4	--	8.9	--	10.1
8	5.0	7.8	--	5.0	5.0	--	7.8	10.1	13.2	15.5	17.0
10	6.6	10.5	--	6.6	6.6	10.5	12.4	15.8	22.6	--	29.5
12	8.7	14.0	--	8.7	13.2	16.4	21.0	27.9	37.2	--	42.7
14 OD	11.5	19.4	11.5	11.5	18.7	23.3	30.3	38.1	52.8	--	--
16 OD	16.4	24.8	16.4	16.4	24.8	29.5	33.5	43.4	58.3	--	--
18 OD	20.1	32.6	20.1	30.3	38.1	54.3	--	--	--	--	--
20 OD	23.3	36.5	23.3	36.5	42.7	65.1	--	--	--	--	--
24 OD	31.0	50.5	31.0	50.5	--	--	--	--	--	--	--

Pipe Thickness: The wall thickness of the pipe determines the man hours that will apply. For swedged ends on double extra strong pipe thickness, use Schedule 160 man hours.

Ends: All man hours are based on ends being furnished either plain or beveled for welding.

Unlisted Sizes: Unlisted sizes take the next higher listing.

ECCENTRIC SWEDGED ENDS

Labor for Welding

Carbon Steel Material

NET MAN HOURS EACH

Size Ins.	Standard Pipe & OD Sizes 3/8" Thick	Extra Heavy Pipe & OD Sizes 1/2" Thick	SCHEDULE NUMBERS								
			20	30	40	60	80	100	120	140	160
2	1.6	1.9	--	--	1.6	--	1.9	--	--	--	3.4
2-1/2	1.7	2.5	--	--	1.7	--	2.5	--	--	--	4.0
3	2.0	2.8	--	--	2.0	--	2.8	--	--	--	4.5
3-1/2	2.3	3.3	--	--	2.3	--	3.3	--	--	--	--
4	2.7	3.9	--	--	2.7	--	3.9	--	6.2	--	6.6
5	3.5	5.2	--	--	3.5	--	5.2	--	8.2	--	9.0
6	4.3	6.2	--	--	4.3	--	6.2	--	11.1	--	12.5
8	6.2	10.1	6.2	6.2	6.2	--	10.1	12.4	17.0	18.7	20.4
10	8.1	13.7	8.1	8.1	8.1	13.7	17.0	21.0	27.2	--	38.8
12	11.6	17.9	11.6	11.6	17.9	21.7	27.9	37.2	48.1	--	54.2
14 OD	16.4	25.6	16.4	16.4	26.3	31.8	38.8	49.7	68.2	--	--
16 OD	23.2	34.1	23.2	23.2	34.1	40.3	44.3	58.3	76.0	--	--
18 OD	27.5	46.5	27.5	27.5	51.2	72.2	--	--	--	--	--
20 OD	30.5	53.3	30.5	30.5	58.3	85.3	--	--	--	--	--
24 OD	43.5	69.8	43.5	43.5	--	--	--	--	--	--	--

Pipe Thickness: The wall thickness of the pipe determines the man hours that will apply. For swedged ends on double extra strong pipe thickness, use Schedule 160 man hours.

Ends: All man hours are based on ends being furnished either plain or beveled for welding.

Unlisted Sizes: Unlisted sizes take the next higher listing.

END CLOSURES

Pressure Type

All Labor

Carbon Steel Materials

NET MAN HOURS EACH

Nom. Pipe Size Ins.	Standard Pipe & OD Sizes 3/8" Thick	Extra Heavy Pipe & OD Sizes 1/2" Thick	SCHEDULE NUMBERS						XX Hy. or 160
			40	60	80	100	120	140	
1-1/2	0.8	0.9	0.8	--	0.8	--	--	--	2.0
2	1.0	1.1	1.0	--	1.1	--	--	--	3.0
2-1/2	1.1	1.3	1.1	--	1.3	--	--	--	3.7
3	1.3	1.6	1.3	--	1.6	--	--	--	3.9
3-1/2	1.4	1.8	1.4	--	1.8	--	--	--	--
4	1.6	2.1	1.6	--	2.1	--	4.9	--	5.2
5	2.0	2.5	2.0	--	2.5	--	6.5	--	6.9
6	2.3	2.9	2.3	--	2.9	--	8.1	--	8.9
8	3.1	4.0	3.1	--	4.0	7.5	10.6	12.5	13.2
10	3.9	5.0	3.9	5.0	8.7	10.8	18.1	--	20.9
12	4.7	6.1	5.7	6.5	11.0	16.0	24.2	19.5	28.6
14	5.6	7.3	6.8	8.4	12.8	17.7	27.9	26.4	39.2
16	6.3	8.2	8.2	10.6	14.1	19.9	32.1	37.7	49.8
18	7.6	9.6	11.8	13.7	18.5	25.1	36.3	47.5	60.4
20	8.2	10.6	14.2	16.8	22.9	30.3	40.5	57.3	--
24	9.0	12.6	16.6	19.9	27.3	35.7	44.2	67.0	--

Pipe Thickness: Wall thickness of pipe determines the man hours that will apply. For double extra strong pipe thickness use Schedule 160 man hours.

Construction: End closures as such are shop fabricated closures; orange peel, saddle, or flat plate type.

Preheating: If specified or required by Codes, add for this operation. See man hours for preheating.

Stress Relieving: Stress relieving of welds in carbon steel material is required by the A.S.A. Code of Pressure Piping where the wall thickness is 3/4" or greater.

All sizes of butt welds shown below the ruled lines are 3/4" or greater in wall thickness and must be stress relieved, if the end closure involves a circumferential weld.

Where stress relieving is required, an extra charge should be made. See man hours for stress relieving.

Unlisted Sizes: Unlisted sizes take the next higher listing.

HEAVY WALL END CLOSURES—PRESSURE TYPE

Carbon Steel Material
NET MAN HOURS EACH

Nominal Pipe Size	WALL THICKNESS IN.							
	.750	1.00	1.25	1.50	1.75	2.00	2.25	2.50
3	4.6	—	—	—	—	—	—	—
4	—	6.9	8.1	9.2	—	—	—	—
5	—	8.4	9.8	11.1	12.5	—	—	—
6	—	13.4	15.7	17.8	20.1	21.9	23.2	—
8	—	14.1	16.5	18.7	21.1	23.0	24.4	26.1
10	—	—	22.3	25.3	28.6	31.2	33.1	35.4
12	—	—	—	35.0	39.6	43.2	45.8	49.0
14	—	—	—	40.4	45.7	49.8	52.8	56.5
16	—	—	—	—	53.2	58.0	61.5	65.7
18	—	—	—	—	—	64.6	68.5	73.2
20	—	—	—	—	—	69.7	73.9	79.0
22	—	—	—	—	—	76.7	81.3	86.9
24	—	—	—	—	—	83.7	88.7	94.8
	2.75	3.00	3.25	3.50	3.75	4.00	4.25	4.50
10	37.8	40.4	—	—	—	—	—	—
12	52.4	56.0	58.6	61.2	—	—	—	—
14	52.8	56.4	59.1	61.8	64.6	67.6	—	—
16	70.2	75.0	78.5	82.0	85.7	89.6	—	—
18	78.3	83.7	87.6	91.5	95.6	99.9	—	—
20	84.5	90.3	94.5	98.8	103.2	107.8	112.1	116.0
22	92.9	99.3	104.0	108.7	113.6	118.7	123.4	127.7
24	101.3	108.3	113.4	118.5	123.8	129.4	134.6	139.3
	4.75	5.00	5.25	5.50	5.75	6.00		
20	120.8	125.0	129.4	133.3	137.3	141.4		
22	132.9	137.6	142.4	146.7	151.1	155.6		
24	145.0	150.1	155.4	160.1	164.9	169.8		

Construction: End closures as such are shop fabricated closures; orange peel, saddle, or flat plate type.

Preheating: If specified or required by code, add for this operation. See man hours for preheating.

Stress Relieving: Stress relieving of welds in carbon steel material is required by the A.S.A. Code of Pressure Piping where the wall thickness is ¾" or greater.

All the above butt welds are ¾" or greater and must be stress relieved, if end closure involves a circumferential weld.

See respective man hour tables for stress relieving.

LARGE O.D. PIPE END CLOSURES—PRESSURE TYPE

Carbon Steel Material
NET MAN HOURS EACH

O.D. Pipe In.	WALL THICKNESS IN.							
	.375	.500	.750	1.00	1.25	1.50	1.75	2.00
26	28.2	34.5	45.1	55.1	64.4	73.4	81.5	89.5
28	29.8	36.4	48.3	59.1	69.1	78.5	87.1	96.1
30	31.3	38.3	50.9	62.3	72.8	82.7	91.7	101.1
32	32.6	39.9	53.0	64.8	75.8	86.1	95.5	105.3
34	33.9	41.5	55.1	67.4	78.8	89.5	99.3	109.5
36	35.1	42.9	57.0	69.7	81.5	92.6	102.7	113.3
38	36.3	44.4	59.0	72.2	84.4	95.9	106.4	117.4
40	37.8	46.2	61.4	75.1	87.8	99.7	110.6	122.0
42	39.3	48.1	63.9	78.1	91.3	103.7	115.0	126.8
44	40.6	49.7	66.0	80.7	94.3	107.1	118.8	131.0
46	42.0	51.4	68.3	83.5	97.6	110.9	123.0	135.7
48	43.4	53.1	70.5	86.2	100.8	114.5	127.0	140.1
54	46.4	56.7	75.3	92.1	107.7	122.3	135.6	149.6
60	49.6	60.7	80.6	98.6	115.3	131.0	145.3	160.3
	2.25	2.50	2.75	3.00	3.25	3.50	3.75	4.00
26	97.2	105.8	113.1	120.3	128.7	134.9	141.4	148.2
28	104.4	113.6	121.4	129.2	138.2	144.8	151.8	159.1
30	109.8	119.5	127.7	135.9	145.4	152.4	159.7	167.4
32	114.4	124.5	133.1	141.6	151.5	158.8	166.4	174.4
34	118.9	129.4	138.3	147.2	157.5	165.1	173.0	181.3
36	123.0	133.8	143.0	152.2	162.9	170.1	178.3	186.9
	4.25	4.50	4.75	5.00	5.25	5.50	5.75	6.00
26	154.3	160.6	167.2	172.9	178.8	184.9	191.2	196.6
28	165.6	172.4	179.5	185.6	191.9	198.4	205.1	210.8
30	174.3	181.4	188.8	195.2	201.8	208.7	215.8	221.8
32	181.6	189.0	196.7	203.4	210.3	217.5	224.9	231.2
34	188.7	196.4	204.5	211.5	218.7	226.1	233.8	240.3
36	194.6	202.6	210.9	218.1	225.5	233.2	241.1	247.9

Construction: End closures as such are shop fabricated closures; orange peel, saddle, or flat plate type.

Preheating: If specified or required by codes, add for this operation. See man hours for preheating.

Stress Relieving: Stress relieving of welds in carbon steel material is required by the A.S.A. Code of Pressure Piping where the wall thickness is ¾″ or greater.

Above wall thickness .750 through 6.00 must be stress relieved, if the end closure involves a circumferential weld.

See respective table for stress relieving.

90° COUPLING WELDS AND SOCKET WELDS

Labor For Cutting And Welding

Carbon Steel Material

NET MAN HOURS EACH

Pipe Sizes Inches	90°—3000# Coupling Weld	90°—6000# Coupling Weld	SOCKET WELDS	
			Sch. 40 & 80 Pipe	Sch. 100 & Heavier Pipe
1/2″ or Less	1.4	1.7	0.5	0.5
3/4	1.6	1.9	0.5	0.6
1	1.8	2.2	0.6	0.7
1-1/4	2.1	2.5	0.8	0.9
1-1/2	2.3	2.8	0.8	1.0
2	2.9	3.6	0.9	1.3
2-1/2	3.4	4.2	1.1	1.4
3	4.0	4.9	1.2	1.7

Man hours shown are for welding of coupling to the O.D. of the pipe only.

If couplings are to be welded to the I.D. of the pipe, add 50% to the above man hours for pipe thickness up to 1 inch, and an additional 12% for each 1/4 inch or fraction thereof of pipe thickness over 1 inch.

Any coupling welded to pipe heavier than schedule 160 should be man houred as a 6,000 pound coupling.

For couplings welded at angles from 45° to less than 90° and couplings attached to fittings increase above man hours 50%.

For couplings welded at angles less than 45% increase above man hours 75%.

Socket welds do not include cut. See respective man hour table for this charge.

'OLET TYPE WELDS

Labor For Cutting And Welding

Carbon Steel Material

NET MAN HOURS EACH

NOMINAL PIPE SIZE		Standard Weight And 2000#	Extra Strong And 3000#	Greater Than Extra Strong And 6000#
Outlet	Header			
1/2	All Sizes	1.3	1.7	2.2
3/4	All Sizes	1.6	1.9	2.6
1	All Sizes	1.8	2.2	2.9
1-1/4	All Sizes	2.0	2.5	3.3
1-1/2	All Sizes	2.5	3.2	4.3
2	All Sizes	3.4	4.2	5.6
2-1/2	All Sizes	4.0	5.1	6.7
3	All Sizes	4.6	5.9	9.2
4	All Sizes	6.1	7.4	9.8
5	All Sizes	6.9	8.1	11.9
6	All Sizes	7.6	8.6	13.9
8	All Sizes	8.4	9.2	16.4
10	All Sizes	11.8	16.9	26.3
12	All Sizes	16.5	19.6	38.9
14	14" and 16"	20.7	23.0	46.9
14	18" And Larger	18.4	20.7	51.0
16	16" and 18"	24.7	26.4	61.2
16	20" and Larger	21.8	23.8	66.3
18	18" and 20"	29.3	32.1	79.1
18	24" and Larger	25.8	28.4	85.2
20	20" and 24"	35.6	39.0	87.8
20	26" and Larger	31.0	34.7	94.6
24	24" and 26"	54.5	63.7	105.3
24	28" and Larger	45.9	55.1	113.5

Man hours are based on the outlet size and schedule except when the run schedule is greater than the outlet schedule, in which case the man hours are based on the outlet size and run schedule.

For elbolet or latrolet welds, and weldolets, threadolets, etc., that are attached to fittings or welded at any angle other than 90°, add 50% to the above man hours.

For sweepolet attachment welds, add 150% to the above man hours.

FLAME CUTTING PIPE—PLAIN ENDS

Labor For Straight Pipe Only

Carbon Steel Material

NET MAN HOURS EACH

Pipe Size Inches	Standard Pipe & O.D. Sizes 3/8" Thick	Extra Hvy. Pipe & O.D. Sizes 1/2" Thick	SCHEDULE NUMBERS								
			20	30	40	60	80	100	120	140	160
2" or Less	0.09	0.13	--	--	0.09	--	0.13	--	--	--	0.18
2-1/2	0.10	0.15	--	--	0.10	--	0.15	--	--	--	0.20
3	0.13	0.18	--	--	0.13	--	0.18	--	--	--	0.24
4	0.18	0.24	--	--	0.18	--	0.24	--	0.33	--	0.36
5	0.21	0.31	--	--	0.21	--	0.31	--	0.38	--	0.43
6	0.29	0.38	--	--	0.29	--	0.38	--	0.49	--	0.55
8	0.40	0.56	0.40	0.40	0.40	0.51	0.56	0.66	0.75	0.84	0.99
10	0.56	0.80	0.56	0.56	0.56	0.80	0.86	0.95	1.08	1.24	1.50
12	0.61	0.95	0.61	0.61	0.75	1.13	1.19	1.29	1.50	1.66	1.78
14 O.D.	0.85	1.13	0.85	0.85	1.00	1.25	1.45	1.55	1.70	2.00	2.10
16 O.D.	0.95	1.40	0.95	0.95	1.40	1.55	1.65	1.85	2.00	2.25	2.55
18 O.D.	1.20	1.70	1.20	1.40	1.70	1.90	2.00	2.25	2.40	2.70	3.15
20 O.D.	1.45	1.90	1.45	1.95	2.10	2.25	2.40	2.65	2.80	3.25	3.70
24 O.D.	2.20	2.80	2.20	2.95	3.10	3.25	3.35	3.65	4.05	4.55	5.15

For mitre cuts less than 30°, add 50% to the above man hours.

For mitre cuts 30° or greater, add 100% to the above man hours.

Man hours are for cutting pipe with plain ends only and do not include beveling, threading, etc. See appropriate man hour tables for these operations and time requirements.

For cutting the ends of bends or trimming fittings, add 50% to the above man hours.

FLAME CUTTING HEAVY WALL PIPE—PLAIN ENDS

Labor For Straight Pipe Only

Carbon Steel Material

NET MAN HOURS EACH

Nominal Pipe Size	WALL THICKNESS IN INCHES							
	.750	1.00	1.25	1.50	1.75	2.00	2.25	2.50
3	0.45	0.80	--	--	--	--	--	--
4	0.80	0.95	1.35	1.50	--	--	--	--
5	--	1.10	1.40	1.70	1.90	2.15	--	--
6	--	1.35	1.60	1.90	2.15	2.35	2.70	--
8	--	1.60	1.90	2.10	2.45	2.65	2.95	3.45
10	--	--	2.10	2.35	2.65	2.90	3.25	3.70
12	--	--	--	2.65	2.85	3.25	3.60	4.05
14	--	--	--	3.05	3.15	3.50	3.90	4.45
16	--	--	--	--	3.65	4.20	4.75	5.20
18	--	--	--	--	--	3.90	5.20	5.80
20	--	--	--	--	--	5.20	5.80	6.55
22	--	--	--	--	--	--	--	7.10
24	--	--	--	--	--	--	--	8.05

	2.75	3.00	3.25	3.50	3.75	4.00	4.25	4.50
10	4.00	4.25	--	--	--	--	--	--
12	4.50	4.85	5.20	5.60	--	--	--	--
14	4.95	5.35	5.60	6.35	6.80	7.45	--	--
16	5.65	6.10	6.55	7.10	7.70	8.45	--	--
18	6.35	6.80	7.45	8.05	8.70	9.65	--	--
20	7.00	7.55	8.30	9.05	9.80	10.65	11.55	12.40
22	7.85	8.50	9.05	10.05	10.95	11.80	12.70	13.65
24	8.70	9.45	10.25	11.20	11.80	13.10	14.30	15.65

	4.75	5.00	5.25	5.50	5.75	6.00		
20	13.25	14.05	15.00	15.80	16.75	17.60		
22	14.60	15.55	16.40	17.30	18.25	19.25		
24	16.90	17.80	18.90	20.00	21.00	22.15		

For mitre cuts less than 30°, add 50% to the above man hours.

For mitre cuts 30° or greater, add 100% to the above man hours.

Man hours are for cutting pipe with plain ends only and do not include beveling, threading, etc. See appropriate man hour tables for these operations and time requirements.

For cutting the ends of bends or trimming fittings, add 50% to the above man hours.

FLAME CUTTING LARGE O. D. PIPE—PLAIN ENDS

Labor For Straight Pipe Only

Carbon Steel Material

NET MAN HOURS EACH

O.D. Pipe Inches	WALL THICKNESS IN INCHES							
	.375	.500	.750	1.00	1.25	1.50	1.75	2.00
26	3.50	4.65	5.25	5.65	5.80	6.10	6.35	6.55
28	4.05	5.05	5.65	6.10	6.30	6.60	6.70	6.95
30	4.30	5.45	6.10	6.35	6.70	6.95	7.15	7.45
32	4.75	5.80	6.40	6.80	7.10	7.45	7.70	7.85
34	5.20	6.40	6.90	7.30	7.55	7.85	8.15	8.35
36	5.80	6.95	7.45	7.85	8.15	8.45	8.70	9.00
38	6.55	7.15	8.05	8.45	8.70	9.10	9.60	9.90
40	7.30	8.15	8.80	9.05	9.45	9.85	10.25	10.65
42	8.15	9.30	9.65	9.90	10.25	10.80	11.20	11.55
44	9.25	10.20	10.65	10.95	11.40	11.75	12.15	12.55
46	10.35	11.20	11.70	11.95	12.35	12.85	13.30	13.65
48	11.70	12.30	12.70	13.10	13.50	13.90	14.45	14.80
54	13.16	13.83	14.28	14.74	15.18	15.63	16.25	16.65
60	14.62	15.37	15.87	16.37	16.87	17.37	18.06	18.50
	2.25	2.50	2.75	3.00	3.25	3.50	3.75	4.00
26	6.90	8.90	9.65	10.40	11.20	12.15	12.70	14.05
28	7.45	9.45	9.85	10.65	11.50	12.55	13.15	14.40
30	7.85	9.90	10.20	10.95	12.00	12.95	13.50	14.85
32	8.30	10.40	10.80	11.40	12.40	13.35	13.90	15.25
34	8.80	10.95	11.35	11.75	12.90	13.75	14.30	15.65
36	9.30	11.40	11.80	12.15	13.30	14.30	14.85	16.20
	4.25	4.50	4.75	5.00	5.25	5.50	5.75	6.00
26	15.25	16.55	17.75	18.90	20.10	21.35	22.55	23.75
28	15.65	17.00	18.15	19.45	20.50	21.75	22.95	24.15
30	16.05	17.30	18.65	19.85	21.00	22.15	23.35	24.55
32	16.55	17.80	19.05	20.25	21.40	22.60	23.75	25.00
34	17.00	18.30	19.50	20.65	21.85	23.10	24.30	25.50
36	17.40	18.75	20.00	21.20	22.40	23.50	24.75	26.05

For mitre cuts less than 30°, add 50% to the above man hours.

For mitre cuts 30° or greater, add 100% to the above man hours.

Man hours are for cutting pipe with plain ends only and do not include beveling, threading, etc. See appropriate man hour tables for these operations and time requirements.

For cutting the ends of bends or trimming fittings, add 50% to the above man hours.

MACHINE CUTTING PIPE—PLAIN ENDS

Labor For Straight Pipe Only

Carbon Steel Material

NET MAN HOURS EACH

Pipe Sizes Inches	Standard Pipe & O.D. Size 3/8" Thick	Extra Hvy. Pipe & O.D. Sizes 1/2" Thick	SCHEDULE NUMBERS									
			20	30	40	60	80	100	120	140	160
2" or Less	0.20	0.29	--	--	0.20	--	0.29	--	--	--	0.40
2-1/2"	0.22	0.33	--	--	0.22	--	0.33	--	--	--	0.44
3	0.29	0.40	--	--	0.29	--	0.40	--	--	--	0.53
4	0.40	0.53	--	--	0.40	--	0.53	--	0.73	--	0.80
5	0.47	0.69	--	--	0.47	--	0.69	--	0.84	--	0.95
6	0.64	0.84	--	--	0.64	--	0.84	--	1.09	--	1.22
8	0.89	1.24	0.89	0.89	0.89	1.13	1.24	1.47	1.67	1.86	2.20
10	1.24	1.78	1.24	1.24	1.24	1.78	1.91	2.11	2.40	2.75	3.33
12	1.35	2.11	1.35	1.35	1.67	2.51	2.64	2.86	3.33	3.69	3.95
14 O.D.	1.89	2.51	1.89	1.89	2.22	2.78	3.22	3.44	3.77	4.44	4.66
16 O.D.	2.11	3.11	2.11	2.11	3.11	3.44	3.66	4.11	4.44	5.00	5.66
18 O.D.	2.66	3.77	2.66	3.11	3.77	4.22	4.44	5.00	5.33	5.99	6.99
20 O.D.	3.22	4.22	3.22	4.33	4.66	5.00	5.33	5.88	6.22	7.22	8.21
24 O.D.	4.88	6.22	4.88	6.55	6.88	7.22	7.44	8.10	8.99	10.10	11.43

For mitre cuts less than 30°, add 50% to the above man hours.

For mitre cuts 30° or greater, add 100% to the above man hours.

Man hours are for cutting pipe with plain ends only and do not include beveling, threading, etc. See appropriate man hours tables for these operations and time requirements.

For cutting the ends of bends or trimming fittings, add 50% to the above man hours.

MACHINE CUTTING HEAVY WALL PIPE—PLAIN ENDS

Labor For Straight Pipe Only

Carbon Steel Material

NET MAN HOURS EACH

Nominal Pipe Size	WALL THICKNESS IN INCHES							
	.750	1.00	1.25	1.50	1.75	2.00	2.25	2.50
3	1.00	1.78	--	--	--	--	--	--
4	1.78	2.11	3.00	3.33	--	--	--	--
5	--	2.44	3.11	3.77	4.22	4.77	--	--
6	--	3.00	3.55	4.22	4.77	5.22	5.99	--
8	--	3.55	4.22	4.66	5.44	5.88	6.55	7.66
10	--	--	4.66	5.22	5.88	6.44	7.22	8.21
12	--	--	--	5.88	6.33	7.22	7.99	8.99
14	--	--	--	6.77	6.99	7.77	8.66	9.88
16	--	--	--	--	8.10	9.32	10.55	11.54
18	--	--	--	--	--	8.66	11.54	12.88
20	--	--	--	--	--	11.54	12.88	14.54
22	--	--	--	--	--	--	--	15.76
24	--	--	--	--	--	--	--	17.87
	2.75	3.00	3.25	3.50	3.75	4.00	4.25	4.50
10	8.88	9.44	--	--	--	--	--	--
12	9.99	10.77	11.54	12.43	--	--	--	--
14	10.99	11.88	12.43	14.10	15.10	16.54	--	--
16	12.54	13.54	14.54	15.76	17.09	18.76	--	--
18	14.10	15.10	16.54	17.87	19.31	21.42	--	--
20	15.54	16.76	18.43	20.09	21.76	23.64	25.64	27.53
22	17.43	18.87	20.09	22.31	24.31	26.20	28.19	30.30
24	19.31	20.98	22.76	24.86	26.20	29.08	31.75	34.74
	4.75	5.00	5.25	5.50	5.75	6.00		
20	49.42	31.19	33.30	35.08	37.19	39.07		
22	32.41	34.52	36.41	38.41	40.52	42.74		
24	37.52	39.52	41.96	44.40	46.62	49.17		

For mitre cuts less than 30°, add 50% to the above man hours.

For mitre cuts 30° or greater, add 100% to the above man hours.

Man hours are for cutting pipe with plain ends only and do not include beveling, threading, etc. see appropriate man hour tables for these operations and time requirements.

For cutting the ends of bends or trimming fittings, add 50% to the above man hours.

MACHINE CUTTING LARGE O.D. PIPE—PLAIN ENDS

Labor For Straight Pipe Only
Carbon Steel Material

NET MAN HOURS EACH

O.D. Pipe Inches	WALL THICKNESS IN INCHES							
	.375	.500	.750	1.00	1.25	1.50	1.75	2.00
26	7.77	10.32	11.66	12.54	12.88	13.54	14.10	14.54
28	8.99	11.21	12.54	13.54	14.10	14.65	14.87	15.43
30	9.55	12.10	13.54	14.10	14.87	15.43	15.87	16.54
32	10.55	12.88	14.21	15.10	15.76	16.54	17.09	17.43
34	11.54	14.21	15.32	16.21	16.76	17.43	18.09	18.54
36	12.88	15.43	16.54	17.43	18.09	18.76	19.31	19.98
38	14.54	15.87	17.87	18.76	19.31	20.20	21.31	21.98
40	16.21	18.09	19.54	20.09	20.98	21.87	22.76	23.64
42	18.09	20.65	21.42	21.98	22.76	23.98	24.86	25.64
44	20.54	22.64	23.64	24.31	25.31	26.09	26.97	27.86
46	22.98	24.86	25.97	26.53	27.42	28.53	29.53	30.30
48	25.97	27.31	28.19	29.08	29.97	30.86	32.08	32.86
54	29.22	30.70	31.70	32.72	33.70	34.70	36.08	36.96
60	32.46	34.12	35.23	36.34	37.45	38.56	40.09	41.07
	2.25	2.50	2.75	3.00	3.25	3.50	3.75	4.00
26	15.32	19.76	21.42	23.09	24.86	26.97	28.19	31.19
28	16.54	20.98	21.87	23.64	25.53	27.86	29.19	31.97
30	17.43	21.98	22.64	24.31	26.64	28.75	29.97	32.97
32	18.43	23.09	23.98	25.31	27.53	29.64	30.86	33.86
34	19.54	24.31	25.20	26.09	28.64	30.53	31.75	34.74
36	20.65	25.31	26.20	26.97	29.52	31.75	32.97	35.96
	4.25	4.50	4.75	5.00	5.25	5.50	5.75	6.00
26	33.86	36.74	39.41	41.96	44.62	47.40	50.06	52.73
28	34.74	37.74	40.29	43.18	45.51	48.29	50.95	53.61
30	35.63	38.41	41.40	44.07	46.62	49.17	51.84	54.50
32	36.74	39.52	42.29	44.96	47.51	50.17	52.73	55.50
34	37.74	40.63	43.29	45.84	48.51	51.28	53.95	56.61
36	38.63	41.63	44.40	47.06	49.73	52.17	54.95	57.83

For mitre cuts less than 30°, add 50% to the above man hours.

For mitre cuts 30° or greater, add 100% to the above man hours.

Man hours are for cutting pipe with plain ends only and do not include beveling, threading, etc. See appropriate man hour tables for these operations and time requirements.

For cutting the ends of bends or trimming fittings, add 50% to the above man hours.

FLAME BEVELING PIPE FOR WELDING

"V" TYPE BEVELS

Labor For Straight Pipe Only
Carbon Steel Material

NET MAN HOURS EACH

Pipe Size Inches	Standard Pipe & O.D. Sizes 3/8" Thick	Extra Hvy. Pipe & O.D. Sizes 1/8" Thick	SCHEDULE NUMBERS								
			20	30	40	60	80	100	120	140	160
2" Or Less	0.07	0.10	--	--	0.07	--	0.10	--	--	--	0.14
2-1/2"	0.08	0.12	--	--	0.08	--	0.12	--	--	--	0.16
3	0.10	0.14	--	--	0.10	--	0.14	--	--	--	0.19
4	0.14	0.19	--	--	0.14	--	0.19	--	0.26	--	0.28
5	0.17	0.24	--	--	0.17	--	0.24	--	0.30	--	0.34
6	0.23	0.30	--	--	0.23	--	0.30	--	0.39	--	0.43
8	0.32	0.44	0.32	0.32	0.32	0.40	0.44	0.52	0.59	0.65	--
10	0.44	0.63	0.44	0.44	0.44	0.63	0.68	0.75	0.83	--	--
12	0.48	0.75	0.48	0.48	0.59	0.89	0.94	1.03	--	--	--
14 O.D.	0.67	0.89	0.67	0.67	0.79	0.98	1.14	--	--	--	--
16 O.D.	0.75	1.10	0.75	0.75	1.10	1.22	1.35	--	--	--	--
18 O.D.	0.94	1.34	0.94	1.10	1.34	1.50	--	--	--	--	--
20 O.D.	1.14	1.50	1.14	1.54	1.65	1.82	--	--	--	--	--
24 O.D.	1.73	2.20	1.73	2.32	2.44	--	--	--	--	--	--

For mitre bevels add 50% to the above man hours.

Above man hours are for flame "V" beveling only and do not include cutting or internal machining. See respective man hour tables for these charges.

For beveling on the ends of bends or shop trimmed fittings, add 50% to the above man hours.

The above man hours are for wall thicknesses of 7/8" or less. For wall thicknesses greater than 7/8" refer to man hours on following pages.

MACHINE BEVELING PIPE FOR WELDING

"U" Type, "V" Type And Double-Angle Bevels

Labor For Straight Pipe Only
Carbon Steel Material

NET MAN HOURS EACH

Pipe Size Inches	Standard Pipe & O.D. Sizes 3/8" Thick	Extra Hvy. Pipe & O.D. Sizes 1/2" Thick	SCHEDULE NUMBERS								
			20	30	40	60	80	100	120	140	160
2" Or Less	0.33	0.36	--	--	0.33	--	0.36	--	--	--	0.87
2-1/2"	0.34	0.37	--	--	0.34	--	0.37	--	--	--	0.91
3	0.36	0.39	--	--	0.36	--	0.39	--	--	--	0.93
4	0.39	0.42	--	--	0.39	--	0.42	--	0.72	--	0.99
5	0.40	0.50	--	--	0.40	--	0.50	--	0.78	--	1.06
6	0.46	0.59	--	--	0.46	--	0.59	--	0.83	--	1.16
8	0.63	0.91	0.63	0.63	0.63	0.81	0.91	1.05	1.18	1.32	1.87
10	0.91	1.26	0.91	0.91	0.91	1.26	1.36	1.50	1.69	2.34	2.83
12	0.97	1.50	0.97	0.97	1.18	1.77	1.87	2.03	2.83	3.14	3.35
14 O.D.	1.34	1.77	1.34	1.34	1.57	1.97	2.28	2.93	3.21	3.78	3.97
16 O.D.	1.50	2.20	1.50	1.50	2.20	2.44	2.60	3.50	3.78	4.25	4.82
18 O.D.	1.89	2.68	1.89	2.20	2.68	2.99	3.21	4.25	4.54	5.10	5.95
20 O.D.	2.28	2.99	2.28	3.07	3.31	3.54	4.54	5.01	5.29	6.14	6.99
24 O.D.	3.46	4.41	3.46	4.65	4.88	6.14	6.33	6.90	7.65	8.60	9.73

For bevels on the ends of bends or shop trimmed fittings, or mitre bevels, add 50% to the above man hours.

For "lip" bevels, add 50% to the above man hours.

For rolled down "lip" bevels, add 75% to the above man hours.

Above man hours are for machine beveling only and do not include cutting or internal machining. See respective man hour tables for these charges.

All pipe sizes shown below the ruled line have a wall thickness greater than 7/8" and must have U-type or double angle bevels in accordance with ANSI and ASME codes for pressure piping. Sizes above the ruled line are 7/8" wall thickness or less. The man hours shown above the ruled line are for bevels as required for inert arc root pass welding.

BEVELING HEAVY WALL PIPE FOR WELDING

"U" Type Or Double Angle Bevels

Labor For Straight Pipe Only
Carbon Steel Material

NET MAN HOURS EACH

Nominal Pipe Size	WALL THICKNESS IN INCHES							
	.750	1.00	1.25	1.50	1.75	2.00	2.25	2.50
3	1.54	1.57	--	--	--	--	--	--
4	1.57	1.61	1.70	1.89	--	--	--	--
5	--	1.73	1.76	2.14	2.39	2.71	--	--
6	--	1.86	2.02	2.39	2.71	2.96	3.40	--
8	--	1.95	2.29	2.65	3.09	3.34	3.72	4.35
10	--	--	2.65	2.96	3.34	3.65	4.09	4.66
12	--	--	--	3.34	3.59	4.09	4.54	5.10
14	--	--	--	3.84	3.97	4.41	4.91	5.61
16	--	--	--	--	4.60	5.29	5.98	6.55
18	--	--	--	--	--	5.98	6.55	7.31
20	--	--	--	--	--	6.55	7.31	8.25
22	--	--	--	--	--	--	--	8.94
24	--	--	--	--	--	--	--	10.14
	2.75	3.00	3.25	3.50	3.75	4.00	4.25	4.50
10	5.04	5.35	--	--	--	--	--	--
12	5.67	6.11	6.55	7.06	--	--	--	--
14	6.24	6.74	7.06	8.00	8.60	9.39	--	--
16	7.12	7.69	8.25	8.94	9.70	10.65	--	--
18	8.00	8.60	9.39	10.14	10.96	12.16	--	--
20	8.82	9.51	10.46	11.40	12.35	13.42	14.55	15.62
22	9.89	10.71	11.40	12.66	13.80	14.87	16.00	17.20
24	10.96	11.91	12.91	14.11	14.87	16.50	18.02	19.72
	4.75	5.00	5.25	5.50	5.75	6.00		
20	16.69	17.70	18.90	19.91	21.10	22.17		
22	18.39	19.59	20.66	21.80	22.99	24.25		
24	21.29	22.43	23.81	25.20	26.46	27.91		

For General Notes, see the bottom of pages 45 and 46.

BEVELING LARGE O. D. PIPE FOR WELDING

Labor For Straight Pipe Only
Carbon Steel Material

NET MAN HOURS EACH

O.D. Pipe Inches	WALL THICKNESS IN INCHES							
	FLAME CUT "V" BEVELS			MACHINE CUT "V" OR DOUBLE ANGLE BEVELS				
	.375	.500	.750	1.00	1.25	1.50	1.75	2.00
26	2.77	3.65	4.13	7.12	7.31	7.69	8.00	8.25
28	3.18	3.97	4.44	7.69	8.00	8.32	8.44	8.76
30	3.40	4.28	4.82	8.00	8.44	8.76	9.01	9.39
32	3.75	4.57	5.04	8.57	8.95	9.39	9.70	9.89
34	4.10	5.04	5.45	9.20	9.51	9.89	10.27	10.52
36	4.57	5.48	5.86	9.89	10.27	10.65	10.96	11.34
38	5.17	6.08	6.33	10.65	10.96	11.47	12.10	12.47
40	5.76	6.65	6.93	11.40	11.91	12.41	12.92	13.42
42	6.43	7.34	7.59	12.47	12.92	13.61	14.11	14.55
44	7.28	8.03	8.38	13.80	14.36	14.81	15.31	15.81
46	8.16	8.82	9.23	15.06	15.56	16.19	16.76	17.20
48	9.23	9.70	10.02	16.51	17.01	17.51	18.21	18.65
54	10.38	10.91	11.28	18.58	19.14	19.70	20.49	20.98
60	11.54	12.13	12.53	20.64	21.26	21.89	22.76	23.31
	2.25	2.50	2.75	3.00	3.25	3.50	3.75	4.00
26	8.69	11.21	12.16	13.10	14.11	15.31	16.00	17.70
28	9.39	11.91	12.41	13.42	14.49	15.81	16.57	18.14
30	9.89	12.47	12.85	13.80	15.12	16.32	17.01	18.71
32	10.46	13.10	13.61	14.36	15.62	16.82	17.51	19.22
34	11.09	13.80	14.30	14.81	16.25	17.33	18.02	19.92
36	11.72	14.36	14.87	15.31	16.76	18.02	18.71	20.41
	4.25	4.50	4.75	5.00	5.25	5.50	5.75	6.00
26	19.22	20.85	22.37	23.81	25.33	26.90	28.41	29.93
28	19.72	21.42	22.87	24.51	25.83	27.41	28.92	30.43
30	20.23	21.80	23.50	25.01	26.46	27.91	29.42	30.93
32	20.85	22.43	24.00	25.52	26.96	28.48	29.93	31.50
34	21.42	23.06	24.57	26.02	27.53	29.11	30.62	32.13
36	21.92	23.63	25.20	26.71	28.22	29.61	31.19	32.82

For General Notes, see the bottom of pages 45 and 46.

THREADING PIPE—INCLUDING CUT

Labor for Cutting and Threading Only
Carbon Steel Material

NET MAN HOURS EACH

Pipe Size Inches	Standard Pipe & O.D. Sizes 3/8" Thick	Extra Hvy. Pipe & O.D. Sizes 1/2" Thick	SCHEDULE NUMBERS								
			20	30	40	60	80	100	120	140	160
2" or less	0.17	0.25	--	--	0.17	--	0.25	--	--	--	0.36
2-1/2	0.23	0.34	--	--	0.23	--	0.34	--	--	--	0.40
3	0.25	0.36	--	--	0.25	--	0.36	--	--	--	0.49
4	0.36	0.49	--	--	0.36	--	0.49	--	0.71	--	0.76
5	0.46	0.66	--	--	0.46	--	0.66	--	0.77	--	0.91
6	0.59	0.79	--	--	0.59	--	0.79	--	1.00	--	1.14
8	0.82	1.14	0.82	0.82	0.82	1.07	1.14	1.39	1.55	1.73	2.02
10	1.17	1.56	1.17	1.17	1.17	1.56	1.79	2.07	2.19	2.61	3.09
12	1.30	2.07	1.30	1.30	1.75	2.33	2.46	2.69	3.09	3.51	3.71
14 O.D.	1.75	2.26	1.75	1.75	2.26	2.50	2.87	3.26	3.51	--	--
16 O.D.	2.01	2.92	2.01	2.01	2.92	3.14	3.51	3.71	4.13	--	--
18 O.D.	2.50	3.51	2.50	3.09	3.51	3.71	4.13	4.59	--	--	--
20 O.D.	3.00	4.00	3.00	4.00	4.81	5.22	5.50	--	--	--	--
24 O.D.	4.39	5.84	4.39	6.13	6.50	6.72	7.09	--	--	--	--

Above man hours are for die cut IPS pipe threads only.

For shop make-on of screwed fittings use 50% of the above man hours.

For threading the ends of bends, add 100% to the above man hours.

WELDED CARBON STEEL ATTACHMENTS

NET MAN HOURS PER LINEAL INCH

Thickness of Plate, etc. Inches	**Layout and Flame Cutting per Lin. Inch		Fillet Welding per Lin. Inch
1/2 or less	....	0.04	....	0.04
3/4	....	0.04	....	0.06
1	....	0.04	....	0.08
1-1/4	....	0.06	....	0.1
1-1/2	....	0.06	....	0.1
1-3/4	....	0.07	....	0.2
2	....	0.08	....	0.2

**Figure labor on basis of total lineal inches to be cut and fillet welded.
Unlisted thicknesses take the next higher listing.
Man hours do not include machining of bases, anchors, supports, lugs, etc.
If preheating is required, add 100% to the above man hours.

DRILLING HOLES IN WELDED ATTACHMENTS

Carbon Steel Material
MAN HOURS EACH

Thickness of Plates, Angles Etc. in Inches	HOLE SIZE			
	3/4" and Smaller	7/8", 1" and 1-1/8"	1-1/4", 1-1/2" and 2"	2-1/4" and 2-1/2"
1/2" or less	0.20	0.24	0.28	0.39
3/4	0.24	0.28	0.36	0.46
1	0.26	0.33	0.41	0.51
1-1/4	0.33	0.41	0.46	0.59
1-1/2	0.41	0.46	0.59	0.76
1-3/4	0.46	0.59	0.72	0.93
2	0.59	0.68	0.84	1.10
2-1/2	0.68	0.73	0.93	1.35
3	0.76	0.93	1.10	1.52
3-1/2	0.84	1.01	1.27	1.78
4	1.01	1.18	1.44	2.03

Unlisted thicknesses of plate or sizes of holes take the next higher listing.
If holes are to be tapped—Add 33-1/3%.
Drilling of Sentinel, Safety or Tell Tale holes will be charged at .05 man hours.
The above man hours are for drilling holes in flat carbon steel plate and structural shapes only.
For drilling holes in pipe or other contoured objects, perpendicular to contoured surface, add 100% to the above man hours.
For drilling holes in pipe or other contoured objects, oblique to contoured surface, add 175% to the above man hours.

MACHINING INSIDE OF PIPE

Built-Up-Ends
Carbon Materials Only

Size Inches	Machining Inside of Pipe Net Man Hours per End		Built Up Ends on Inside Diameter of Pipe and Fittings with Weld Metal to Provide for Specified Outside Diameter of Machined Backing Ring	
	Standard Extra Strong & Sch. Nos. to 100 Inclusive	Double Extra Strong & Sch. Nos. 120, 140 & 160	Size Inches	Net Man Hours per End
2 or less	0.4	0.6	2 or less	0.5
2-1/2	0.4	0.6	2-1/2	0.5
3	0.4	0.6	3	0.6
3-1/2	0.4	0.7	3-1/2	0.6
4	0.6	0.7	4	0.7
5	0.7	0.8	5	0.8
6	0.7	0.9	6	0.9
8	0.9	1.1	8	1.2
10	1.0	1.3	10	1.7
12	1.1	1.5	12	2.1
14 OD	1.3	1.8	14	2.6
16 OD	1.5	2.1	16	3.2
18 OD	1.8	2.4	18	3.9
20 OD	2.1	2.9	20	4.7
24 OD	2.9	3.8	24	7.1

Machining: Man hours for machining the inside of straight pipe are for any taper bore from 10° through 30° included angle. For machining the ends of bends add 100% to the above man hours. For counterboring (up to a maximum of 2" in length), add 50% to the above man hours. For machining to a controlled "C" dimension (as required for power piping critical systems), add 225% to the above man hours.

Cutting and Beveling: Man hours do not include cutting and beveling. See respective tables for these charges.

Built-Up Ends: Man hours for built-up ends are for building up the I.D. of straight pipe, bends or fittings, at the ends with weld metal and grinding where it is necessary for proper fit of backing rings.

MACHINING INSIDE OF LARGE O.D. PIPE

Built-Up Ends
Carbon Steel Material

O.D. Pipe Size Inches	NET MAN HOURS PER END Machining Inside of Straight Pipe Only					I.D. Build-Up with Weld Metal
	WALL THICKNESS IN INCHES					
	.500 to 1.50	1.51 to 2.25	2.26 to 3.00	3.01 to 4.50	4.51 to 6.00	Man Hours Per End
26	3.74	4.49	5.35	6.84	8.57	12.88
28	4.03	4.83	5.75	7.25	9.03	15.24
30	4.49	5.18	6.15	7.71	9.55	19.26
32	4.95	5.75	6.50	8.22	10.18	23.58
34	5.58	6.27	7.13	8.80	10.70	29.67
36	6.27	7.02	7.76	9.32	11.27	35.31
38	7.02	7.82	8.63	10.00	11.90	41.40
40	7.82	8.63	9.55	10.70	12.59	48.36
42	8.68	9.37	10.47	11.39	13.34	56.70
44	9.49	10.41	11.39	12.25	14.03	65.67
46	10.41	11.27	12.36	13.23	14.89	74.00
48	11.39	12.25	13.28	14.15	15.76	83.43
54	12.81	13.78	14.94	15.92	17.73	93.86
60	14.24	15.31	16.60	17.69	19.70	104.29

Machining: Man hours for machining the inside of straight pipe are for any taper bore from 10° through 30° included angle. For machining the ends of bends add 100% to the above man hours. For counterboring (up to a maximum of 2" in length), add 50% to the above man hours. For machining to a controlled "C" dimension (as required for power piping critical systems), add 225% to the above man hours.

Cutting and Beveling: Man hours do not include cutting and beveling. See respective tables for these charges.

Built-Up Ends: Man hours for built-up ends are for building up the I.D. of straight pipe, bends or fittings, at the ends with weld metal and grinding where it is necessary for proper fit of backing rings.

BORING INSIDE DIAMETER OF PIPE
AND INSTALLING STRAIGHTENING VANES

NET MAN HOURS EACH

Nominal Pipe Size Inches	Boring I.D. of Pipe	Installing Straightening Vanes	
	Carbon Steel	Carbon Steel	Alloy
4	8.3	6.4	9.6
5	9.9	7.4	11.1
6	11.3	9.1	12.9
8	14.8	10.7	16.0
10	17.7	11.8	17.7
12	21.7	13.2	20.1
14	25.0	14.9	22.3
16	30.0	16.5	25.0
18	37.3	18.7	28.0
20	48.9	21.0	31.5
24	67.0	25.6	38.5
26	--	30.4	45.8
28	--	33.6	50.7
30	--	38.9	58.3
32	--	45.0	67.7
34	--	50.7	76.2
36	--	58.3	87.5
38	--	65.2	98.1
40	--	72.2	108.2
42	--	79.4	119.2

Man hours for boring I.D. only include boring pipe for a length of four times nominal pipe size.

Man hours for installing straightening vanes are based on installing vanes in pipe where boring the I.D. of pipe is not required. If boring I.D. of pipe is required or specified, add boring man hours as shown above.

INSTALLING FLOW NOZZLES
Holding Ring Type

Carbon Steel and Alloy Materials

NET MAN HOURS EACH

Pipe Size Inches	Flow Nozzles		Pipe O.D. Inches	Flow Nozzles	
	Carbon Steel	Alloy		Carbon Steel	Alloy
4	32.9	38.4	26	140.3	168.4
5	35.7	41.4	28	160.7	188.1
6	39.8	46.1	30	184.2	211.8
8	46.8	53.0	32	210.6	239.2
10	53.4	61.6	34	240.4	268.0
12	60.2	70.0	36	270.4	302.9
14 O.D.	65.8	77.3	38	303.0	342.4
16 O.D.	74.2	87.6	40	339.4	386.9
18 O.D.	83.7	99.3	42	380.1	437.3
20 O.D.	94.1	113.3	--	--	--
24 O.D.	118.9	144.7	--	--	--

Man hours include internal machining and nozzle installation.

For installing welding type flow nozzles, add for the bevels, butt weld, butt weld preheat, and any other labor operation or non-destructive testing operation required for the butt weld. See respective tables for these charges.

PREHEATING BUTT WELDS AND ANY TYPE OF FLANGE WELDS

Labor Only

Carbon Steel, or Alloy Materials
For Temperatures Up To 400°F.

NET MAN HOURS EACH

Size Ins.	Standard Pipe & OD Sizes 3/8" Thick	Extra Heavy Pipe & OD Sizes 1/2" Thick	SCHEDULE NUMBERS								
			20	30	40	60	80	100	120	140	160
2	0.2	0.3	--	--	0.2	--	0.3	--	--	--	0.4
2-1/2	0.3	0.4	--	--	0.3	--	0.4	--	--	--	0.5
3	0.4	0.5	--	--	0.4	--	0.5	--	--	--	0.6
3-1/2	0.4	0.5	--	--	0.4	--	0.5	--	--	--	0.8
4	0.5	0.6	--	--	0.5	--	0.6	--	0.8	--	0.8
5	0.6	0.8	--	--	0.6	--	0.8	--	0.8	--	0.9
6	0.7	0.9	--	--	0.7	--	0.9	--	1.1	--	1.3
8	0.8	1.1	0.8	0.8	0.8	1.1	1.1	1.5	1.6	2.0	2.1
10	1.1	1.5	1.1	1.1	1.1	1.5	1.7	2.0	2.3	2.8	3.2
12	1.3	1.7	1.3	1.3	1.6	1.9	2.4	2.8	3.2	3.7	4.5
14 OD	1.6	2.1	1.6	1.6	1.9	2.5	3.0	3.7	4.2	4.9	5.6
16 OD	1.9	2.8	1.9	1.9	2.5	3.2	3.8	4.6	5.1	6.2	7.2
18 OD	2.2	3.0	2.2	2.6	3.5	4.2	5.1	5.9	6.7	7.2	8.9
20 OD	2.6	3.5	2.6	3.5	4.4	5.3	6.3	7.4	8.3	9.4	10.9
24 OD	3.1	4.2	3.1	4.5	5.4	6.6	7.9	8.8	9.9	11.3	12.9

Pipe Thickness: The wall thickness of the material determines the man hours that will apply. For preheating of double extra strong material, use Schedule 160 man hours.

Mitre Welds: For preheating of mitre welds, add 50% to above man hours.

Man Hours: Man hours for preheating are additional to charges for welding operations.

Preheating: For preheating to temperatures above 400°F. but not exceeding 600°F., add 100% to the above man hours.

PREHEATING HEAVY WALL PIPE BUTT WELDS
LABOR ONLY

Carbon Steel or Alloy Materials
For Temperatures Up to 400°F.

NET MAN HOURS EACH

Nominal Pipe Size	WALL THICKNESS IN INCHES							
	.750	1.00	1.25	1.50	1.75	2.00	2.25	2.50
3	0.9	1.0	--	--	--	--	--	--
4	1.2	1.3	1.5	1.7	--	--	--	--
5	--	1.6	1.8	2.0	2.1	2.4	--	--
6	--	1.8	2.1	2.3	2.5	2.7	2.9	--
8	--	2.5	2.9	3.1	3.3	3.7	3.8	4.1
10	--	--	3.5	3.7	4.0	4.6	4.9	5.3
12	--	--	--	5.2	5.6	5.9	6.3	6.8
14	--	--	--	6.2	6.7	7.1	7.7	8.1
16	--	--	--	--	8.0	8.5	8.9	9.8
18	--	--	--	--	--	10.4	11.0	11.6
20	--	--	--	--	--	11.9	12.8	13.5
22	--	--	--	--	--	--	--	14.6
24	--	--	--	--	--	--	--	15.8
	2.75	3.00	3.25	3.50	3.75	4.00	4.25	4.50
10	6.0	6.4	--	--	--	--	--	--
12	7.3	7.7	8.3	8.8	--	--	--	--
14	8.7	9.2	9.8	10.4	11.0	11.9	--	--
16	10.4	11.0	11.6	12.4	13.2	14.1	--	--
18	12.5	13.4	14.2	15.0	15.9	16.8	--	--
20	14.4	15.4	16.4	17.4	18.5	19.4	20.3	21.3
22	15.6	16.8	17.7	18.9	20.2	21.4	22.5	23.7
24	16.8	18.0	19.3	20.5	21.9	23.5	24.1	25.6
	4.75	5.00	5.25	5.50	5.75	6.00		
20	22.4	23.5	24.8	26.0	27.2	28.6		
22	24.8	26.0	27.2	28.6	29.7	30.9		
24	27.2	28.6	29.7	30.9	32.2	33.5		

For General Notes, see the bottom of page 55.

PREHEATING LARGE O.D. PIPE BUTT WELDS
AND ANY TYPE FLANGE WELDS

Carbon Steel Material
For Temperatures Up to 400 °F.

NET MAN HOURS EACH

O.D. Pipe Inches	WALL THICKNESS IN INCHES							
	.500 or less	.750	1.00	1.25	1.50	1.75	2.00	2.25
26	6.4	7.2	8.7	10.6	12.3	13.6	15.5	17.4
28	7.0	7.8	9.2	11.1	13.0	14.4	16.5	18.6
30	7.5	8.1	10.0	11.9	13.8	15.3	17.6	19.9
32	7.9	8.7	10.4	12.5	14.6	16.2	18.6	21.0
34	8.5	9.2	11.0	13.4	15.8	17.4	19.7	22.5
36	9.1	10.0	11.7	14.6	17.4	19.1	21.8	24.6
38	9.2	10.6	12.7	15.5	19.3	21.0	24.0	27.0
40	9.5	11.4	13.8	16.5	21.4	23.0	26.3	29.8
42	10.2	12.2	14.9	17.8	23.8	25.4	28.9	32.7
44	11.0	13.0	16.5	20.3	24.8	27.9	31.9	36.7
46	11.8	13.9	18.0	22.3	26.7	30.6	34.8	39.1
48	12.7	14.9	19.5	24.2	28.7	33.3	38.0	42.6
54	14.3	16.8	21.9	27.2	32.3	37.5	42.7	47.9
60	15.9	18.6	24.4	30.2	35.9	41.6	47.5	53.3

	2.50	2.75	3.00	3.25	3.50	3.75	4.00	4.25
26	19.3	21.1	23.0	24.9	26.8	28.6	30.5	32.4
28	20.5	22.3	24.6	26.1	28.0	29.9	31.8	33.7
30	21.8	23.7	25.7	27.2	29.2	31.0	33.3	34.8
32	22.9	24.8	27.0	28.4	30.3	32.2	34.4	36.0
34	24.4	26.3	28.4	29.9	31.9	33.7	35.9	37.5
36	25.6	28.4	30.5	32.1	34.0	35.9	38.0	39.7

	4.50	4.75	5.00	5.25	5.50	5.75	6.00	
26	34.3	36.2	38.0	39.9	41.8	43.7	45.6	
28	35.6	37.3	39.2	41.1	43.0	44.9	47.2	
30	36.7	38.6	40.5	42.4	44.1	46.0	48.3	
32	37.8	39.7	41.6	43.5	45.4	47.3	49.5	
34	39.3	40.7	43.1	45.0	46.6	48.7	51.0	
36	41.5	43.4	45.3	47.2	48.9	50.8	53.1	

For General Notes, see the bottom of page 55.

PREHEATING 90° NOZZLE WELDS

Carbon Steel, or Alloy Materials
For Temperatures Up to 400°F

NET MAN HOURS EACH

Size Ins.	Standard Pipe & OD Sizes 3/8" Thick	Extra Heavy Pipe & OD Sizes 1/2" Thick	SCHEDULE NUMBERS								
			20	30	40	60	80	100	120	140	160
2	0.4	0.5	--	--	0.4	--	0.5	--	--	--	0.6
2-1/2	0.5	0.6	--	--	0.5	--	0.6	--	--	--	0.8
3	0.5	0.8	--	--	0.5	--	0.8	--	--	--	0.9
3-1/2	0.6	0.8	--	--	0.6	--	0.8	--	--	--	--
4	0.8	0.9	--	--	0.8	--	0.9	--	1.2	--	1.4
5	0.9	1.2	--	--	0.9	--	1.2	--	1.4	--	1.6
6	1.1	1.6	--	--	1.1	--	1.6	--	1.7	--	2.1
8	1.4	1.8	1.4	1.4	1.4	1.7	1.8	2.3	2.6	3.0	3.5
10	1.7	2.3	1.7	1.7	1.7	2.3	2.6	3.2	3.8	4.4	5.0
12	2.1	2.8	2.1	2.1	2.4	3.0	3.8	4.4	5.1	5.9	7.0
14 OD	2.5	3.2	2.5	2.5	3.0	3.9	5.0	5.6	6.6	7.3	8.9
16 OD	2.9	3.9	2.9	2.9	3.9	5.0	6.1	7.2	8.3	9.1	11.6
18 OD	3.6	4.7	3.6	4.1	5.3	6.7	8.1	8.9	10.8	11.4	14.4
20 OD	4.7	5.5	4.2	5.5	7.0	8.6	10.0	11.6	13.2	13.5	15.5
24 OD	5.1	6.6	5.1	7.2	8.7	10.6	12.8	14.0	16.0	16.7	18.6

Pipe Thickness: The size of the nozzle and the wall thickness of the header or nozzle (whichever is greater) determines the man hours to be used. For preheating of double extra strong thickness use schedule 160 man hours.

Time: For reinforced 90° nozzle welds, add 100% to the above man hours. For 45° nozzle welds, add 50% to the above man hours. For reinforced 45° nozzle welds, add 150% to the above man hours. For preheating to temperatures above 400°F. but not exceeding 600°F., add 100% to the above man hours. Preheating of coupling, weldolet, threadolet or socket welds should be charged at the same man hours as shown for the same size and schedule nozzle. Man hours for preheating are additional to man hours for welding operations.

PREHEATING LARGE O.D. 90° NOZZLE WELDS

Carbon Steel, or Alloy Materials
For Temperatures Up to 400°F.

NET MAN HOURS EACH

O.D. Pipe Sizes·	WALL THICKNESS IN INCHES							
	.500	.750	1.00	1.25	1.50	1.75	2.00	2.25
26	8.2	9.1	10.9	13.2	15.5	17.0	19.5	21.9
28	8.7	9.7	11.7	14.1	16.3	18.0	20.7	23.5
30	9.2	10.3	12.5	14.9	17.4	19.3	22.1	25.0
32	9.8	10.9	13.1	15.7	18.3	20.3	23.3	26.3
34	10.7	11.7	13.8	16.9	19.7	21.8	24.8	28.3
36	11.4	12.5	14.8	18.3	21.8	23.9	27.3	30.9
38	12.1	13.4	16.0	19.5	24.2	26.3	30.1	34.0
40	13.0	14.4	17.2	20.8	26.9	29.0	33.1	37.4
42	13.7	15.3	18.5	22.3	28.8	32.0	36.4	41.1
48	15.7	17.5	21.1	25.5	32.9	36.6	41.6	47.0
54	17.6	19.7	23.8	28.7	37.0	41.1	46.8	52.8
60	19.6	21.9	26.4	31.9	41.1	45.7	52.0	58.7

Pipe Thickness: The size of the nozzle and the wall thickness of the header or nozzle (whichever is greater) determines the man hours to be used.

Time: For reinforced 90° nozzle welds, add 100% to the above man hours.
For 45° nozzle welds, add 50% to the above man hours.
For reinforced 45° nozzle welds, add 150% to the above man hours.
For preheating to temperatures above 400°F. but not exceeding 600°F., add 100% to the above man hours.
Preheating of coupling, weldolet, threadolet or socket welds should be charged at the same man hours as shown for the same size and schedule nozzle.
Man hours for preheating are additional to man hours for welding operations.

LOCAL STRESS RELIEVING

Gas or Electric Method—Butt Welds—Nozzle Welds or Any Type of Flange Welds
Carbon Steel Materials
Temperatures to 1400°F.

NET MAN HOURS EACH

Size Ins.	Standard Pipe & OD Sizes 3/8"Thick	Extra Heavy Pipe & OD Sizes 1/2"Thick	SCHEDULE NUMBERS								
			20	30	40	60	80	100	120	140	160
2	2.3	2.4	--	--	2.3	--	2.4	--	--	--	2.6
2-1/2	2.4	2.5	--	--	2.4	--	2.5	--	--	--	2.7
3	2.5	2.6	--	--	2.5	--	2.6	--	--	--	3.0
3-1/2	2.6	2.7	--	--	2.6	--	2.7	--	--	--	3.3
4	2.6	3.0	--	--	2.6	--	3.0	--	3.1	--	3.4
5	3.0	3.2	--	--	3.0	--	3.2	--	3.5	--	3.7
6	3.2	3.6	--	--	3.2	--	3.6	--	3.7	--	4.3
8	3.6	4.0	3.6	3.6	3.6	3.7	4.0	4.3	4.5	4.7	5.0
10	3.9	4.3	3.9	3.9	3.9	4.3	4.5	4.8	5.0	5.3	5.7
12	4.3	4.7	4.3	4.3	4.5	4.9	5.1	5.5	5.8	6.0	6.3
14 OD	4.7	5.0	4.7	4.7	5.0	5.3	5.7	6.0	6.4	6.7	7.0
16 OD	5.0	5.4	5.0	5.0	5.4	5.8	6.1	6.6	6.8	7.2	7.8
18 OD	5.4	5.8	5.4	5.6	5.8	6.2	6.6	7.0	7.4	7.8	8.6
20 OD	5.6	5.9	5.6	5.8	6.2	6.6	7.0	7.8	8.1	8.5	9.4
24 OD	6.0	6.2	6.0	6.4	6.8	7.2	7.8	8.6	8.9	9.5	10.6

Pipe Thickness: For Stress relieving butt welds and flange welds, the wall thickness of the pipe determines the man hours that will apply. For stress relieving nozzle welds, the size and thickness of the header to which the nozzle is attached determines the man hours that will apply. For local stress relieving of double extra strong material, use Schedule 160 man hours.

Man Hours: The total man hours for stress relieving shall be determined as follows:
(1) By computing the total of all welds contained in the complete requirement figure on the basis of local stress relieving unit man hours;
(2) By totaling all pieces included in the complete requirement which can be full furnace stress relieved as units, classifying them in their applicable groups, and computing the total man hours.
Whichever of these two methods develops the lower man hours should be used in determining the man hours for stress relieving.

Valves: Stress relieving may be done by the local stress relieving process, or, unless valves have been welded into the assembly, the complete fabricated assembly may be full furnace stress relieved as a unit.

The stress relieving of butt welds joining valves to fabricated assemblies must be man houred as the man hours covering local stress relieving, even though adjacent assemblies can be full furnace stress relieved as a unit.

Code Requirements: All welds in piping materials having a wall thickness of 3/4" or greater must be stress relieved to comply with the requirements of the A.S.A. Code for Pressure Piping. Man hours shown below the ruled line in the above schedule cover sizes having a wall thickness of 3/4" or greater.

HEAVY WALL LOCAL STRESS RELIEVING
BUTT WELDS

Carbon Steel Material
Temperatures To 1400° F.

NET MAN HOURS EACH

Nominal Pipe Size	WALL THICKNESS IN INCHES							
	.750	1.00	1.25	1.50	1.75	2.00	2.25	2.50
3	4.5	4.7	--	--	--	--	--	--
4	4.7	5.2	5.6	6.0	--	--	--	--
5	--	5.5	5.9	6.3	6.7	7.2	--	--
6	--	5.9	6.3	6.8	7.2	7.9	8.5	--
8	--	6.5	6.8	7.4	7.8	8.3	8.9	9.5
10	--	--	7.1	7.7	8.1	8.6	9.0	9.8
12	--	--	--	7.9	8.4	9.0	9.5	10.1
14	--	--	--	8.4	9.0	9.8	10.2	10.8
16	--	--	--	--	9.4	10.1	10.7	11.4
18	--	--	--	--	--	10.7	11.4	12.2
20	--	--	--	--	--	11.6	12.4	13.2
22	--	--	--	--	--	--	--	14.3
24	--	--	--	--	--	--	--	15.4
	2.75	3.00	3.25	3.50	3.75	4.00	4.25	4.50
10	10.4	11.0	--	--	--	--	--	--
12	10.8	11.6	12.3	13.2	--	--	--	--
14	11.6	12.4	13.2	14.1	14.8	15.9	--	--
16	12.2	13.0	13.9	14.7	15.7	16.8	--	--
18	13.0	13.9	14.7	15.9	16.9	18.1	--	--
20	14.1	15.1	16.0	17.0	18.4	19.7	21.1	22.5
22	15.1	16.7	17.4	18.5	19.8	21.1	22.5	23.8
24	16.4	17.5	18.7	19.9	21.3	22.6	24.0	25.3
	4.75	5.00	5.25	5.50	5.75	6.00		
20	23.8	25.2	26.5	27.9	29.3	30.7		
22	25.2	26.6	27.9	29.3	30.7	32.1		
24	26.6	28.0	29.3	30.7	32.0	33.4		

For General Notes, see the bottom of page 60.

LARGE O.D. LOCAL STRESS RELIEVING
Butt Welds, Nozzle Welds or Any Type of Flange Weld

Carbon Steel Material
Temperatures to 1400° F.

NET MAN HOURS EACH

O.D. Pipe Size	WALL THICKNESS IN INCHES								
	.375	.500	.750	1.00	1.25	1.50	1.75	2.00	2.25
26	7.9	8.5	9.8	10.3	11.0	12.3	14.2	16.1	18.5
28	8.3	9.1	10.6	11.0	12.0	13.2	15.4	17.4	19.9
30	9.0	9.8	11.4	12.1	12.8	14.5	16.7	18.7	21.4
32	9.8	10.7	12.4	13.0	13.7	15.7	18.1	20.2	23.0
34	10.8	11.7	13.5	14.1	14.9	17.0	19.6	21.6	24.5
36	11.8	12.8	14.8	15.7	16.8	18.7	21.3	23.5	26.6
38	12.8	14.0	16.3	17.4	18.8	20.8	23.5	25.5	29.1
40	14.0	15.3	17.9	19.4	21.1	23.1	25.8	27.8	31.7
42	15.5	16.9	19.7	21.4	23.7	25.7	28.3	30.4	34.5
44	17.3	18.7	21.5	23.5	25.8	28.3	31.1	34.1	37.2
46	19.0	20.5	23.4	26.0	28.4	31.3	34.2	37.2	40.4
48	21.2	22.6	25.4	28.4	31.1	34.2	37.3	40.3	43.3
54	24.8	25.4	28.6	32.0	35.0	38.5	42.0	45.3	48.7
60	26.4	28.2	31.8	35.5	38.9	42.8	46.6	50.4	55.3
	2.50	2.75	3.00	3.25	3.50	3.75	4.00	4.25	4.50
26	20.7	23.0	25.3	27.6	29.9	32.2	34.5	36.8	39.1
28	22.2	24.5	26.8	29.1	31.4	33.7	36.0	38.3	40.5
30	23.7	26.0	28.3	30.6	32.9	35.1	37.4	39.7	42.0
32	25.3	27.6	29.8	32.1	34.4	36.7	39.0	41.3	43.6
34	26.8	29.1	31.4	33.0	36.0	38.3	40.6	42.9	45.2
36	28.9	31.2	32.4	35.1	38.1	40.4	42.7	45.0	47.3
	4.75	5.00	5.25	5.50	5.75	6.00			
26	41.4	43.7	45.9	48.2	50.5	52.8			
28	42.8	45.1	47.4	49.7	52.0	54.3			
30	44.3	46.6	48.9	51.1	53.4	55.7			
32	45.9	48.2	50.5	52.7	55.0	57.3			
34	47.5	49.8	52.1	54.3	56.6	58.9			
36	49.9	51.8	54.1	56.4	58.7	61.0			

For General Notes, see the bottom of page 60.

FULL FURNACE STRESS RELIEVING AND HEATING TREATMENT

Carbon Steel and Alloy Materials

NET MAN HOURS

Fahrenheit Temperature	Per Hundred Pounds
0° to 1250° Inclusive	0.3
1251° to 1400° Inclusive.	0.4
1401° to 1700° Inclusive.	0.7
1701° to 2200° Inclusive.	1.1

Exposed sections of pieces too large to be placed entirely within the furnace will be included in subsequent furnace heat or heats until all parts of the piece have been stress relieved or heat treated. To calculate the man hours for this operation use: total weight of fabricated piece times man hours per pound depending on temperature, times total number of times piece must be heated to get full coverage.

Quenching is included in the above man hours. Materials to be quenched after annealing must not exceed over-all furnace dimension.

RADIOGRAPHIC INSPECTION
X-Ray or Gamma Ray Inspection of Butt Welds

Carbon Steel Material

NET MAN HOURS EACH

Nominal Pipe Size	Wall Thickness Thru Extra Strong	Wall Thickness Greater Than Extra Strong Thru Schedule 120	Wall Thickness Greater Than Schedule 120 Thru Double Extra Strong
2 or less	0.75	--	0.98
3	0.75	--	0.98
4	0.85	0.98	1.10
5	0.93	1.07	1.20
6	1.04	1.20	1.36
8	1.17	1.34	1.52
10	1.31	1.50	1.71
12	1.49	1.71	1.94
14	1.62	1.86	2.10
16	1.81	2.08	2.35
18	2.02	2.32	2.62
20	2.22	2.56	2.90
24	2.74	3.15	3.55

Man hours listed above cover radiographic inspection of butt welded joints by X-raying or gamma-ray, at the option of the client.

For radiographic inspection of mitre butt welds add 50% to above man hours.

For radiographic inspection of slip-on flange welds add 100% to above man hours.

For radiographic inspection of nozzle welds add 200% to above man hours.

HEAVY WALL RADIOGRAPHIC INSPECTION
X-Ray or Gamma Ray Inspection of Butt Welds

Carbon Steel Material

NET MAN HOURS EACH

Nominal Pipe Size	WALL THICKNESS IN INCHES							
	.750	1.00	1.25	1.50	1.75	2.00	2.25	2.50
3	1.06	1.17	--	--	--	--	--	--
4	1.17	1.22	1.25	1.41	--	--	--	--
5	--	1.28	1.36	1.47	1.55	1.68	--	--
6	--	1.36	1.47	1.55	1.68	1.79	1.92	--
8	--	1.49	1.60	1.71	1.85	1.95	2.10	2.35
10	--	--	1.78	1.87	2.05	2.14	2.30	2.53
12	--	--	--	2.03	2.21	2.34	2.51	2.67
14	--	--	--	2.24	2.38	2.58	2.74	2.91
16	--	--	--	--	2.59	2.77	2.94	3.15
18	--	--	--	--	--	3.02	3.25	3.44
20	--	--	--	--	--	3.31	3.50	3.76
22	--	--	--	--	--	--	--	4.19
24	--	--	--	--	--	--	--	4.62
	2.75	3.00	3.25	3.50	3.75	4.00	4.25	4.50
10	2.70	2.85	--	--	--	--	--	--
12	2.85	3.07	3.25	3.47	--	--	--	--
14	3.10	3.31	3.50	3.76	4.00	4.27	--	--
16	3.39	3.60	3.81	4.11	4.37	4.62	--	--
18	3.70	3.95	4.21	4.45	4.77	5.12	--	--
20	4.00	4.27	4.54	4.83	5.16	5.50	5.87	6.22
22	4.46	4.77	5.10	5.41	5.76	6.19	6.61	7.01
24	4.94	5.30	5.66	5.98	6.42	6.85	7.49	7.79
	4.75	5.00	5.25	5.50	5.75	6.00		
20	6.61	6.98	7.33	7.70	8.03	8.40		
22	7.30	7.65	8.00	8.35	8.72	9.07		
24	8.08	8.34	8.66	9.06	9.44	9.82		

For General Notes, see the bottom of page 64.

LARGE O.D. RADIOGRAPHIC INSPECTION
X-Ray or Gamma Ray Inspection of Butt Welds

Carbon Steel Material

NET MAN HOURS EACH

O.D. Pipe Size	WALL THICKNESS IN INCHES							
	.750 or Less	1.00	1.25	1.50	1.75	2.00	2.25	2.50
26	3.15	3.31	3.44	3.71	3.97	4.27	4.56	4.90
28	3.65	3.78	3.94	4.19	4.46	4.75	5.07	5.42
30	4.40	4.56	4.69	4.94	5.23	5.50	5.79	6.18
32	5.44	5.57	5.70	5.98	6.27	6.54	6.80	7.17
34	6.74	6.90	7.02	7.30	7.57	7.84	8.13	8.50
36	8.29	8.45	8.59	8.86	9.14	9.42	9.73	10.05
38	9.89	10.06	10.24	10.50	10.78	11.06	11.38	--
40	11.62	11.86	12.03	12.27	12.61	12.91	13.22	--
42	13.50	13.70	13.89	14.19	14.54	14.82	15.14	--
44	15.54	15.70	15.92	16.22	16.59	16.91	17.22	--
46	17.63	17.84	18.08	18.40	18.77	19.07	19.36	--
48	19.86	20.06	20.29	20.61	21.04	21.30	21.62	--
54	22.34	22.57	22.83	23.18	23.67	23.96	24.32	--
60	24.82	25.07	25.36	25.76	26.30	26.62	27.02	--

	2.75	3.00	3.25	3.50	3.75	4.00	4.25	4.50
26	5.31	5.81	6.22	6.56	7.06	7.49	8.32	8.70
28	5.84	6.54	6.78	7.09	7.58	8.10	8.86	9.22
30	6.56	7.33	7.52	7.84	8.34	8.86	9.58	9.95
32	7.57	8.32	8.51	8.83	9.33	9.86	10.59	10.94
34	8.90	9.63	9.86	10.16	10.66	11.17	11.92	12.29
36	10.48	11.22	11.42	11.73	12.22	12.75	13.50	13.87

	4.75	5.00	5.25	5.50	5.75	6.00		
26	9.00	9.26	9.52	9.92	10.18	10.56		
28	9.52	9.81	10.05	10.43	10.69	11.09		
30	10.29	10.54	10.80	11.17	11.46	11.84		
32	11.28	11.54	11.79	12.16	13.15	13.57		
34	12.59	12.88	13.10	13.50	14.50	15.26		
36	14.18	14.45	14.69	15.07	16.06	15.85		

For General Notes, see the bottom of page 64.

MAGNETIC OR DYE PENETRANT INSPECTION OF WELDED JOINTS

All Thicknesses and Schedules

NET MAN HOURS EACH

Size Inches	MAGNETIC		DYE PENETRANT	
	Butt Welds	Nozzle Welds	Butt Welds	Nozzle Welds
2 or less	0.6	0.9	0.8	1.3
2-1/2	0.7	1.1	0.9	1.4
3	0.8	1.2	1.1	1.6
3-1/2	0.9	1.4	1.2	1.9
4	1.1	1.6	1.5	2.2
5	1.4	2.1	1.9	2.8
6	1.7	2.6	2.3	3.5
8	2.0	3.0	2.8	4.1
10	2.5	3.7	3.3	5.0
12	3.1	4.7	4.2	6.3
14 OD	3.4	5.2	4.6	6.9
16 OD	3.9	5.8	5.2	7.9
18 OD	4.4	6.6	5.8	8.7
20 OD	4.8	7.2	6.5	9.7
24 OD	5.6	8.5	7.5	11.1

Man hours above are for a single inspection. When specifications call for multiple inspections during the progress of welding, the man hours shown above will apply for each of the total number of inspections.

Magnetic particle or liquid penetrant inspection of weld end preparations should be charged at the same man hours as comparable inspection of the same size butt weld.

For inspection of reinforced nozzle welds, add 150% to the above man hours to include both the nozzle weld and the pad weld.

For inspection of slip-on flange welds add 50% to the above man hours.

For inspection of small connections such as couplings, bosses, thredolets and weldolets use the man hours shown for corresponding sizes of nozzle welds.

MAGNETIC OR DYE PENETRANT INSPECTION OF WELDED JOINTS

All Thicknesses and Schedules

NET MAN HOURS EACH

O.D. Pipe Size	ALL THICKNESSES AND SCHEDULES			
	Magnetic		Dye Penetrant	
	Butt Welds	Nozzle Welds	Butt Welds	Nozzle Welds
26	5.9	8.8	7.8	11.7
28	6.4	9.5	8.5	12.6
30	6.9	10.3	9.2	13.7
32	7.3	10.9	9.7	14.5
34	7.8	11.6	10.4	15.4
36	8.2	12.4	10.9	16.5
38	8.8	13.1	11.7	17.4
40	9.3	13.9	12.4	18.5
42	9.8	14.7	13.0	19.6
44	10.3	15.5	13.7	20.6
46	10.8	16.3	14.4	21.7
48	11.4	17.0	15.2	22.6
54	12.8	19.1	17.0	25.4
60	14.3	21.2	19.0	28.2

Man hours above are for a single inspection. When specifications call for multiple inspections during the progress of welding, the man hours shown above will apply for each of the total number of inspections.

Magnetic particle or liquid penetrant inspection of weld end preparations should be charged at the same man hours as comparable inspection of the same size butt weld.

For inspection of reinforced nozzle welds, add 150% to the above man hours to include both the nozzle weld and the pad weld.

For inspection of slip-on flange welds add 50% to the above man hours.

For inspection of small connections such as couplings, bosses, thredolets and weldolets use the man hours shown for corresponding sizes of nozzle welds.

TESTING FABRICATED ASSEMBLIES
Hydrostatic Testing of Flanged Ends

Carbon Steel Material

For Pressures Not Exceeding 4,000 P.S.I.

NET MAN HOURS PER FLANGED OUTLET

Nominal Pipe Size	300 Lb. or Less	400 Lb. and 600 Lb.	900 Lb. and 1500 Lb.	2500 Lb.
2" or less	1.0	1.2	1.3	1.6
2-1/2	1.3	1.5	1.6	2.0
3	1.4	1.6	1.8	2.1
4	1.6	1.8	2.0	2.5
5	1.8	2.0	2.2	2.7
6	2.1	2.2	2.7	3.0
8	2.7	3.0	3.3	3.9
10	3.2	3.7	4.2	5.1
12	4.2	4.5	5.2	6.8
14	5.0	5.3	6.1	--
16	6.3	6.4	7.5	--
18	7.2	8.1	9.1	--
20	8.5	9.3	10.7	--
24	12.6	14.1	15.7	--

Above man hours are for flanged ends only. See following tables for plain or beveled ends.

Man hours are for a maximum holding time of one hour at test pressure.

TESTING FABRICATED ASSEMBLIES
Hydrostatic Testing of Plain or Beveled Ends Only

Carbon Steel Material
For Pressures Not Exceeding 4,000 P.S.I.

NET MAN HOURS PER END

Pipe Size Inches	Standard Pipe & O.D. Sizes 3/8" Thick	Extra Hvy. Pipe & O.D. Sizes 1/2" Thick	SCHEDULE NUMBERS								
			20	30	40	60	80	100	120	140	160
2" or Less	2.5	2.8	--	--	2.5	--	2.8	--	--	--	4.4
2-1/2"	2.6	3.1	--	--	2.6	--	3.1	--	--	--	5.0
3	3.0	3.4	--	--	3.0	--	3.4	--	--	--	5.6
4	3.5	4.2	--	--	3.5	--	4.2	--	5.7	--	6.4
5	3.9	4.6	--	--	3.9	--	4.6	--	6.2	--	7.4
6	4.4	5.1	--	--	4.4	--	5.1	--	7.0	--	8.4
8	4.9	5.8	4.9	4.9	4.9	5.3	5.8	7.0	8.2	9.2	10.1
10	5.5	6.3	5.5	5.1	5.5	6.3	7.3	8.7	9.8	11.2	12.5
12	6.1	7.0	6.1	6.1	6.7	7.9	9.5	10.9	12.2	13.7	15.6
14 O.D.	6.8	7.8	6.8	6.8	7.4	9.0	11.0	12.4	14.0	16.2	18.9
16 O.D.	7.8	8.8	7.8	7.8	8.8	11.1	13.4	15.4	17.3	20.4	23.9
18 O.D.	9.1	10.4	9.1	9.8	11.5	13.7	16.9	19.2	21.9	25.1	29.3
20 O.D.	10.1	11.9	10.1	11.9	14.0	16.9	20.6	23.8	26.9	30.7	36.2
24 O.D.	13.5	15.4	13.5	15.6	20.6	25.1	31.3	35.8	41.1	47.3	54.3

Above man hours are for plain or beveled ends only. See preceding table for flanged ends.

Man hours are for a maximum holding time of one hour at test pressure.

HEAVY WALL TESTING FABRICATED ASSEMBLIES
Hydrostatic Testing of Plain or Beveled Ends Only

Carbon Steel Material

For Pressures Not Exceeding 4,000 P.S.I.

NET MAN HOURS PER END

Nominal Pipe Size	WALL THICKNESS IN INCHES							
	.750	1.00	1.25	1.50	1.75	2.00	2.25	2.50
3″ or less	6.3	6.9	--	--	--	--	--	--
4	7.2	7.5	7.9	8.5	--	--	--	--
5	--	8.2	8.7	9.2	10.5	12.9	--	--
6	--	9.5	10.3	10.8	13.9	16.3	19.0	--
8	--	11.5	12.1	13.6	17.3	20.1	23.7	28.6
10	--	--	13.5	16.9	21.0	24.3	28.5	33.5
12	--	--	--	20.5	24.3	28.6	33.9	39.2
14	--	--	--	23.4	27.7	32.5	38.1	45.0
16	--	--	--	--	31.1	37.2	43.4	51.9
18	--	--	--	--	--	41.5	48.4	57.2
20	--	--	--	--	--	48.4	57.2	69.2
22	--	--	--	--	--	--	--	75.6
24	--	--	--	--	--	--	--	82.2
	2.75	3.00	3.25	3.50	3.75	4.00	4.25	4.50
10	38.3	44.0	--	--	--	--	--	--
12	44.8	51.3	57.8	66.0	--	--	--	--
14	51.1	58.0	65.7	74.4	84.8	95.2	--	--
16	58.8	67.5	76.2	86.5	98.7	112.5	--	--
18	65.7	76.2	86.5	99.3	113.3	129.7	--	--
20	78.8	88.3	101.2	114.2	129.7	147.1	166.8	180.9
22	86.0	96.9	110.3	124.6	141.7	161.0	182.5	202.0
24	93.5	103.4	119.5	135.1	154.0	174.9	198.3	218.7
	4.75	5.00	5.25	5.50	5.75	6.00		
20	198.3	212.4	226.6	236.0	262.7	280.1		
22	214.0	229.7	251.4	264.0	289.2	308.3		
24	233.5	251.4	274.7	287.9	311.5	333.2		

Above man hours are for plain or beveled ends only. See preceeding table for flanged ends.

Man hours are for a maximum holding time of one hour at test pressure.

ACCESS HOLES

Carbon Steel Material

NET MAN HOURS EACH

Nominal Pipe Size	WALL THICKNESS				
	Up to 1"	Over 1" to 2"	Over 2" to 2-1/2"	Over 2-1/2" to 4"	Over 4" to 6"
2-1/2, 3, 4	1.6	1.7	--	--	--
5, 6, 8	1.8	1.9	2.3	--	--
10, 12	1.9	2.1	2.5	3.5	--
14, 16, 18	2.0	2.2	2.6	3.7	--
20, 22, 24	2.2	2.5	2.7	3.9	5.9
26, 28, 30	2.5	2.8	2.9	4.2	6.3
32, 34, 36	2.7	2.9	3.2	4.4	6.6
38, 40, 42	2.9	3.2	3.5	4.9	7.3
44, 46, 48	3.2	3.5	3.8	5.8	8.8
54, 60	4.0	4.3	4.7	7.2	10.9

Man hours include access holes through 1" diameter (drilled and tapped) for radiographic inspection of welds when specified or required.

For openings larger than 1" in diameter add 25% to the above man hours for each 1/4" increase in diameter.

If plugs are to be included and seal welded, add 0.5 man hours each.

MISCELLANEOUS FABRICATION OPERATIONS

Descaling R. T. J. Flange Faces: Oxidation (scale) created in R. T. J. grooves of flanges because of Stress Relieving or Heat Treating should be removed and charged for at the following man hours.

Flange Size Inches	Man Hours	Flange Size Inches	Man Hours
3 or less	0.6	14	2.5
4	0.9	16	3.2
6	1.0	18	3.5
8	1.3	20	3.9
10	1.6	24	4.6
12	1.9	--	--

Miscellaneous X-Rays: If specified, the following X-Rays should be taken and charged accordingly. Slip-on Welds X-rayed will be charged the same man hours as a Butt Weld X-ray.

Mitre and Nozzle Welds X-rayed should be charged at 50% more than the Butt Weld X-ray man hours.

Lineal Welding X-rayed should be charged at 0.7 man hours per foot through 1" thickness and at 1.0 man hours per foot for thickness greater than 1". For alloys add 25% to these man hours.

Preheating Coupling Welds: On any size you should charge the man hours of Preheating a 2" extra heavy Nozzle Weld.

MAN HOURS PER FOOT OF CYLINDRICAL COIL
FABRICATION BENDING ONLY

"A" — Coils containing 40 Ft. and Less
"B" — Coils containing 40 Ft. to 100 Ft.
"C" — Coils containing 100 Ft. and More

Diameter of Coil (C-C)	1/2" PIPE						3/4" PIPE					
	Schedule 10-60			Schedule 80-160			Schedule 10-60			Schedule 80-160		
	A	B	C	A	B	C	A	B	C	A	B	C
18" to 36"	0.06	0.05	0.04	0.06	0.05	0.04	0.06	0.05	0.04	0.06	0.05	0.05
36" to 60"	0.05	0.04	0.04	0.05	0.04	0.04	0.06	0.04	0.04	0.06	0.05	0.04
60" and over	0.05	0.04	0.04	0.05	0.04	0.04	0.05	0.04	0.04	0.05	0.04	0.04

Diameter of Coil (C-C)	1" PIPE						1-1/4" PIPE					
	Schedule 10-60			Schedule 80-160			Schedule 10-60			Schedule 80-160		
	A	B	C	A	B	C	A	B	C	A	B	C
18" to 36"	0.07	0.06	0.05	0.07	0.06	0.05	0.08	0.06	0.06	0.08	0.07	0.06
36" to 60"	0.07	0.05	0.05	0.07	0.05	0.05	0.07	0.06	0.05	0.08	0.06	0.06
60" and over	0.06	0.05	0.04	0.06	0.05	0.05	0.07	0.06	0.05	0.07	0.06	0.05

Diameter of Coil (C-C)	1-1/2" PIPE						2" PIPE					
	Schedule 10-60			Schedule 80-160			Schedule 10-60			Schedule 80-160		
	A	B	C	A	B	C	A	B	C	A	B	C
18" to 36"	0.09	0.08	0.07	0.10	0.08	0.08	0.12	0.09	0.08	0.13	0.10	0.09
36" to 60"	0.09	0.07	0.06	0.09	0.08	0.07	0.09	0.09	0.08	0.11	0.10	0.08
60" and over	0.08	0.07	0.06	0.09	0.07	0.07	0.09	0.08	0.07	0.11	0.09	0.08

Work Included: Man hours include bending only. All welding, handling and erection are additional. See respective pages for these man hours.

Section Two

FIELD FABRICATION AND ERECTION

This section is intended to suffice for the complete labor involved in the installation and field fabrication as may be necessary to put a system of process piping into operation in an industrial or chemical plant.

The man hours listed are for labor only and do not have any bearing on materials or equipment.

All labor for unloading from railroad cars or trucks hauling to and unloading at storage facilities, hauling from storage to erection site and rigging or hoisting into place have been given due consideration in the man hours listed for the various operations. While it is true that the aforementioned operations involve more time than is required merely to haul materials from and on the job, fabricating shop or storage area, nevertheless, we have found that these are operations that will balance themselves out over a complete piping job. No consideration has been given to overhead or profit.

For the field fabrication and erection of alloy and non-ferrous piping and fittings, apply the percentages which appear under Section Three to the following pages listing the various field operations.

HANDLING AND ERECTING STRAIGHT RUN PIPE

DIRECT MAN HOURS — PER FOOT

Pipe Size Inches	SCHEDULE NUMBERS		
	10 to 60	80 to 100	120 to 160
1/4	0.16	0.17	0.18
3/8	0.16	0.17	0.19
1/2	0.16	0.18	0.20
3/4	0.17	0.19	0.21
1	0.17	0.20	0.23
1-1/4	0.18	0.21	0.24
1-1/2	0.19	0.22	0.27
2	0.20	0.24	0.29
2-1/2	0.21	0.26	0.32
3	0.23	0.28	0.35
3-1/2	0.24	0.30	0.38
4	0.25	0.31	0.39
5	0.26	0.34	0.43
6	0.28	0.38	0.50
8	0.34	0.48	0.65
10	0.43	0.60	0.82
12	0.52	0.73	1.00
14 OD	0.64	0.87	1.19
16 OD	0.75	1.02	1.39
18 OD	0.88	1.17	1.60
20 OD	1.03	1.32	1.81
24 OD	1.15	1.49	2.04

Man hours include all labor for unloading and storing in yard, loading and hauling to erection site, and rigging and aligning in place. It does not include welding, bolt-ups, make-ons or scaffolding. See respective pages for these items.

For brass, copper, and everdur pipe, double above man hours.

HANDLING AND ERECTING HEAVY WALL STRAIGHT RUN PIPE

Carbon Steel Material

NET MAN HOURS PER FOOT

Nominal Pipe Size	WALL THICKNESS IN INCHES							
	.750	1.00	1.25	1.50	1.75	2.00	2.25	2.50
3″ or less	0.36	0.43	--	--	--	--	--	--
4	0.40	0.48	0.56	0.65	--	--	--	--
5	--	0.54	0.65	0.76	0.87	0.98	--	--
6	--	0.60	0.71	0.81	1.09	1.19	1.30	--
8	--	0.71	0.80	0.95	1.10	1.25	1.40	1.57
10	--	--	0.95	1.00	1.18	1.36	1.54	1.72
12	--	--	--	1.17	1.34	1.51	1.69	1.86
14	--	--	--	1.35	1.50	1.66	1.81	1.97
16	--	--	--	--	1.56	1.73	1.89	2.07
18	--	--	--	--	--	1.78	1.96	2.15
20	--	--	--	--	--	1.80	1.98	2.22
22	--	--	--	--	--	--	--	2.28
24	--	--	--	--	--	--	--	2.33

	2.75	3.00	3.25	3.50	3.75	4.00	4.25	4.50
10	1.98	2.24	--	--	--	--	--	--
12	2.14	2.42	2.73	3.09	--	--	--	--
14	2.26	2.56	2.89	3.27	3.67	4.07	--	--
16	2.38	2.69	3.04	3.43	3.84	4.26	--	--
18	2.47	2.79	3.16	3.57	3.99	4.43	--	--
20	2.55	2.89	3.27	3.69	4.13	4.59	5.09	5.60
22	2.62	2.96	3.34	3.78	4.23	4.70	5.21	5.73
24	2.71	3.06	3.46	3.91	4.37	4.86	5.39	5.92

	4.75	5.00	5.25	5.50	5.75	6.00		
20	6.16	6.71	7.25	7.55	8.38	8.97		
22	6.31	6.88	7.43	8.02	8.58	9.18		
24	6.52	7.10	7.67	8.29	8.87	9.49		

Man hours include all labor for unloading and storing in yard, loading and hauling to erection site, and rigging and aligning in place.

Man hours do not include welding, bolt-ups, make-ons or scaffolding. See respective tables for these items.

HANDLING AND ERECTING LARGE O.D.
STRAIGHT RUN PIPE

Carbon Steel Material

NET MAN HOURS PER FOOT

O.D. Pipe Inches	WALL THICKNESS IN INCHES							
	.500 or less	.750	1.00	1.25	1.50	1.75	2.00	2.25
26	1.18	1.24	1.45	1.54	1.61	1.94	2.03	2.17
28	1.33	1.40	1.54	1.71	1.87	2.01	2.17	2.33
30	1.42	1.49	1.63	1.79	1.92	2.14	2.39	2.60
32	1.52	1.60	1.77	1.93	2.13	2.30	2.49	2.66
34	1.61	1.70	1.87	2.05	2.26	2.44	2.66	2.83
36	1.78	1.88	2.05	2.23	2.40	2.67	2.98	3.23
38	1.88	1.98	2.16	2.35	2.53	2.82	3.15	3.41
40	1.98	2.09	2.28	2.48	2.67	2.97	3.31	3.59
42	2.15	2.26	2.46	2.68	2.88	3.20	3.58	--
44	2.25	2.37	2.58	2.81	3.02	3.35	3.75	--
46	2.36	2.47	2.70	2.93	3.16	3.51	3.92	--
48	2.51	2.64	2.88	3.14	3.35	3.73	4.13	--
54	2.87	3.02	3.29	3.59	3.83	4.26	4.72	--
60	3.23	3.40	3.71	4.04	4.31	4.78	5.31	--

	2.50	2.75	3.00	3.25	3.50	3.75	4.00	4.25
26	2.41	2.78	3.08	3.49	3.92	4.40	4.91	5.41
28	2.47	2.84	3.16	3.54	3.96	4.46	4.96	5.45
30	2.68	2.89	3.21	3.59	4.00	4.52	5.05	5.56
32	2.74	2.93	3.24	3.65	4.06	4.60	5.12	5.67
34	2.92	3.06	3.31	3.73	4.11	4.69	5.20	5.79
36	3.33	3.49	3.67	3.89	4.20	4.74	5.29	5.97

	4.50	4.75	5.00	5.25	5.50	5.75	6.00	
26	5.94	6.54	7.15	7.73	8.33	8.93	9.60	
28	6.05	6.72	7.40	7.85	8.48	9.07	9.68	
30	6.12	7.01	7.63	7.99	8.64	9.35	9.96	
32	6.20	7.20	7.84	8.06	8.80	9.70	10.39	
34	6.38	7.42	8.06	8.32	8.96	9.83	10.57	
36	6.53	7.84	8.27	8.48	9.22	10.04	11.13	

Man hours include all labor for unloading and storing in yard, loading and hauling to erection site, and rigging and aligning in place.

Man hours do not include welding, bolt-ups, make-ons or scaffolding. See respective tables for these items.

HANDLING AND ERECTING FABRICATED
SPOOL PIECES

Carbon Steel Material

DIRECT MAN HOURS — PER FOOT BY SIZE

Pipe Size Inches	SCHEDULE NUMBERS		
	10 to 60	80 to 100	120 to 160
1/4	0.26	0.29	0.30
3/8	0.27	0.29	0.32
1/2	0.27	0.30	0.34
3/4	0.28	0.32	0.35
1	0.29	0.34	0.39
1-1/4	0.30	0.35	0.41
1-1/2	0.32	0.37	0.45
2	0.34	0.40	0.49
2-1/2	0.36	0.44	0.54
3	0.39	0.48	0.59
3-1/2	0.40	0.50	0.62
4	0.41	0.52	0.66
5	0.44	0.57	0.72
6	0.47	0.64	0.84
8	0.57	0.81	0.99
10	0.72	1.00	1.38
12	0.88	1.23	1.69
14 OD	1.01	1.46	2.01
16 OD	1.27	1.71	2.34
18 OD	1.48	1.96	2.69
20 OD	1.74	2.22	3.04
24 OD	1.94	2.51	3.43

Man hours are for labor only and includes handling and hauling from storage yard, unloading and rigging in place, and aligning. It does not include welding, bolt-ups, make-ons or scaffolding. See other pages for these charges.

For brass, copper and everdur pipe, double above man hours.

Units apply to any length spool piece or segment of work.

HANDLING AND ERECTING HEAVY WALL FABRICATED SPOOL PIECES

Carbon Steel Material

	NET MAN HOURS PER FOOT							
Nominal Pipe Size	WALL THICKNESS IN INCHES							
	.750	1.00	1.25	1.50	1.75	2.00	2.25	2.50
3″ or less	0.61	0.67	--	--	--	--	--	--
4	0.68	0.72	0.84	0.98	--	--	--	--
5	--	0.81	0.98	1.14	1.31	1.47	--	--
6	--	0.90	1.07	1.22	1.64	1.79	1.95	--
8	--	1.07	1.20	1.43	1.65	1.88	2.10	2.36
10	--	--	1.43	1.50	1.77	2.04	2.31	2.58
12	--	--	--	1.76	2.01	2.27	2.54	2.79
14	--	--	--	2.03	2.25	2.49	2.72	2.96
16	--	--	--	--	2.42	2.60	2.84	3.11
18	--	--	--	--	--	2.76	2.94	3.23
20	--	--	--	--	--	3.07	3.17	3.33
22	--	--	--	--	--	--	--	3.42
24	--	--	--	--	--	--	--	3.50
	2.75	3.00	3.25	3.50	3.75	4.00	4.25	4.50
10	2.77	3.14	--	--	--	--	--	--
12	3.00	3.39	3.82	4.33	--	--	--	--
14	3.16	3.58	4.05	4.58	5.14	5.70	--	--
16	3.33	3.77	4.26	4.80	5.38	5.96	--	--
18	3.46	3.91	4.42	5.00	5.59	6.20	--	--
20	3.57	4.05	4.58	5.17	5.78	6.43	7.13	7.84
22	3.67	4.14	4.68	5.29	5.92	6.58	7.29	8.02
24	3.79	4.28	4.84	5.47	6.12	6.80	7.55	8.29
	4.75	5.00	5.25	5.50	5.75	6.00		
20	8.01	8.72	9.43	9.82	10.89	11.66		
22	8.20	8.94	9.66	10.43	11.15	11.93		
24	8.48	9.23	9.97	10.78	11.53	12.34		

Man hours are for labor only and include handling and hauling from storage yard, unloading and rigging in place, and aligning. This does not include welding, bolt-ups, make-ons or scaffolding. See other pages for these charges.

Units apply to any length spool piece or segment of work.

HANDLING AND ERECTING LARGE O.D. FABRICATED SPOOL PIECES

Carbon Steel Material

NET MAN HOURS PER FOOT

O.D. Pipe Inches	WALL THICKNESS IN INCHES							
	.500 or less	.750	1.00	1.25	1.50	1.75	2.00	2.25
26	1.77	1.86	2.18	2.31	2.42	2.91	3.05	3.26
28	1.97	2.07	2.28	2.53	2.77	2.97	3.21	3.45
30	2.07	2.18	2.38	2.61	2.80	3.12	3.49	3.80
32	2.17	2.29	2.53	2.76	3.05	3.29	3.56	3.86
34	2.25	2.38	2.62	2.87	3.16	3.42	3.72	3.96
36	2.46	2.59	2.83	3.08	3.31	3.68	4.11	4.46
38	2.56	2.69	2.94	3.20	3.44	3.84	4.28	4.64
40	2.65	2.80	3.06	3.32	3.58	3.98	4.44	4.81
42	2.84	2.98	3.25	3.54	3.80	4.22	4.73	--
44	2.93	3.08	3.35	3.65	3.93	4.36	4.88	--
46	3.02	3.16	3.46	3.75	4.04	4.49	5.02	--
48	3.19	3.35	3.66	3.99	4.25	4.74	5.25	--
54	3.64	3.84	4.18	4.60	4.86	5.41	5.99	--
60	4.07	4.28	4.67	5.09	5.43	6.02	6.69	--
	2.50	2.75	3.00	3.25	3.50	3.75	4.00	4.25
26	3.62	4.17	4.62	5.24	5.88	6.60	7.37	8.12
28	3.66	4.20	4.68	5.30	5.94	6.67	7.43	8.18
30	3.91	4.27	4.75	5.35	6.00	6.73	7.52	8.28
32	4.08	4.37	4.83	5.44	6.05	6.85	7.63	8.45
34	4.35	4.60	4.93	5.56	6.12	6.99	7.75	8.63
36	4.96	5.20	5.47	5.68	6.26	7.06	7.88	8.90
	4.50	4.75	5.00	5.25	5.50	5.75	6.00	
26	8.85	9.74	10.66	11.52	12.41	13.31	14.30	
28	9.01	10.01	11.03	11.70	12.64	13.51	14.42	
30	9.12	10.44	11.37	11.91	12.87	13.93	14.84	
32	9.24	10.73	11.68	12.00	13.11	14.45	15.48	
34	9.51	11.06	12.00	12.40	13.35	14.65	15.75	
36	9.73	11.68	12.32	12.64	13.74	14.96	16.58	

Man hours are for labor only and include handling and hauling from storage yard, unloading and rigging in place, and aligning. This does not include welding, bolt-ups, make-ons or scaffolding. See other pages for these charges.

Units apply to any length spool piece or segment of work.

MAKING ON SCREWED FITTINGS AND VALVES

NET MAN HOURS EACH

Nom. Size Inches	PER CONNECTION	
	PLAIN	BACK WELDED
1/4	0.1	0.4
3/8	0.1	0.4
1/2	0.1	0.4
3/4	0.1	0.5
1	0.2	0.5
1-1/4	0.2	0.6
1-1/2	0.3	0.7
2	0.3	0.9
2-1/2	0.4	1.0
3	0.4	1.2
3-1/2	0.4	1.4
4	0.5	1.6
6	.7	2.3
8	.9	2.8
10	1.1	3.4
12	1.2	3.9
14	1.3	4.2
16	1.4	4.5
18	1.5	4.8
20	1.6	5.1
24	1.7	5.5

Man hours per connection only. For cutting, threading, field handling and erection, additional man hours are required. See pages pertaining to these operations.

Ells and Valves = Two Connections
Tees   = Three Connections
Crosses   = Four Connections

FIELD HANDLING VALVES

NET MAN HOURS EACH

Pipe Size Inches	SERVICE PRESSURE RATING				
	150 Lb.	300-400 Lb.	600-900 Lb.	1500 Lb.	2500 Lb.
¼	0.2	0.2	0.2	0.4	0.4
⅜	0.2	0.2	0.3	0.4	0.5
½	0.2	0.2	0.3	0.5	0.5
¾	0.2	0.3	0.5	0.6	0.6
1	0.3	0.3	0.6	0.6	0.7
1¼	0.3	0.3	0.7	0.9	1.2
1½	0.4	0.4	1.0	1.2	1.4
2	0.5	0.8	1.3	1.5	1.8
2½	0.8	1.1	1.5	1.9	2.1
3	1.2	1.5	2.0	2.4	2.6
3½	1.4	1.7	2.3	2.7	2.9
4	1.7	2.0	2.6	3.1	3.4
5	2.0	2.4	3.0	3.6	4.0
6	2.2	2.7	3.3	4.1	4.1
8	2.8	3.4	4.2	5.3	5.9
10	3.6	4.2	5.1	6.8	7.0
12	4.3	5.1	6.3	8.5	9.5
14	5.1	6.0	7.5	10.5	11.4
16	5.9	7.1	8.8	12.7	13.0
18	6.7	8.1	10.4	15.1	15.8
20	7.7	9.2	11.9	17.9	18.4
24	8.5	10.3	13.6	20.7	21.5
26	8.9	10.8	14.2	21.7	—
28	9.3	11.3	14.9	22.7	—
30	9.7	11.8	15.5	23.6	—
32	10.1	12.3	16.1	24.6	—
34	10.5	12.8	16.8	25.6	—
36	10.9	13.2	17.4	26.5	—
38	11.3	13.7	18.0	27.4	—
40	11.6	14.1	18.5	28.2	—
42	11.9	14.5	19.1	29.0	—
44	12.3	14.9	19.7	29.9	—
46	12.7	15.3	20.3	30.8	—
48	13.0	15.7	20.8	31.5	—

Man hours only — screwed, flanged, and weld end valves, and expansion joints. No man hours for welds, making-on, or bolt-up included. See pages pertaining to these items.

Use 150# allowance for standard brass and iron valves.

Use 300# allowance for extra heavy and 200 lb. brass and iron valves.

For motor operated or diaphragm valves, add 125% to above man hours.

FIELD ERECTION BOLT-UPS

NET MAN HOURS EACH

Pipe Size Inches	SERVICE PRESSURE RATING					
	150 Lb.	300-400 Lb.	600 Lb.	900 Lb.	1500 Lb.	2500 Lb.
2 or less	0.7	0.8	0.9	1.0	1.2	1.6
2½	0.8	0.9	1.0	1.2	1.5	2.0
3	0.8	0.9	1.0	1.2	1.5	2.0
3½	1.0	1.2	1.3	1.5	1.8	2.4
4	1.2	1.4	1.5	1.7	2.1	2.8
6	1.5	1.7	1.8	2.1	2.6	3.4
8	2.1	2.4	2.6	3.0	3.7	4.9
10	2.7	3.0	3.2	3.7	4.6	6.1
12	3.4	3.8	4.1	4.7	5.8	7.7
14	3.8	4.3	4.6	5.3	6.5	—
16	4.4	4.9	5.2	6.0	7.4	—
18	4.8	5.4	5.8	6.7	8.2	—
20	5.5	6.2	6.6	7.6	9.3	—
24	6.6	7.4	7.9	9.1	11.2	—
26	7.0	7.8	8.4	9.6	—	—
28	7.4	8.3	8.9	10.2	—	—
30	7.8	8.7	9.4	10.7	—	—
32	8.2	9.2	9.9	11.3	—	—
34	8.6	9.6	10.3	11.8	—	—
36	9.0	10.0	10.8	12.3	—	—
38	9.4	10.4	11.3	12.8	—	—
40	9.7	10.8	11.7	13.3	—	—
42	10.1	11.2	12.1	13.8	—	—

Man hours for labor only for each joint on valves, flanged fittings, and spools. Above man hours do not include handling of valves, fittings or spools. The handling of bolts or studs and gaskets *is* included.

Where tongue and groove, ring joint, female or fittings with special facings are used, add 25% to above units.

For standard cast iron use 150# allowance.

For extra heavy cast iron use 300# allowance.

ATTACHING FLANGES – SCREWED TYPE

Man hours — Cutting and Threading Pipe — Making on Screwed Flanges and Refacing

Carbon Steel Material for Bends, Headers,
Necks and Straight Runs of Pipe

NET MAN HOURS EACH

Pipe Size Inches	125 Lb. Cast Iron and 150 Lb. Steel	250 Lb. Cast Iron and Steel 300 Lb. and Higher
2 or less	1.2	1.4
2-1/2	1.3	1.5
3	1.4	1.6
3-1/2	1.6	1.8
4	1.7	2.0
5	1.8	2.2
6	2.1	2.3
8	2.5	2.8
10	3.1	3.4
12	3.7	4.1
14 OD	4.5	5.1
16 OD	5.4	6.1
18 OD	6.5	7.3
20 OD	7.7	8.7
24 OD	11.0	12.5

Flanges: Man hours are for field labor only. The price of the flange must be added in all cases.

Pipe Thickness: Man hours are for any wall thickness of pipe used with listed flanges.

Unlisted Sizes: Unlisted sizes take the next higher listing.

ATTACHING FLANGES—SCREWED TYPE

Man Hours—Cutting and Threading Pipe, Making on Flange
Manual Seal Welding at Back and Front and Refacing

Welded or Seamless Carbon Steel Material, Straight Pipe,
Bends, Headers and Nozzles

NET MAN HOURS EACH

Pipe Size Inches	SERVICE PRESSURE RATING						
	150 Lb.	300 Lb.	400 Lb.	600 Lb.	900 Lb.	1500 Lb.	2500 Lb.
2 or less	2.0	2.2	2.4	2.4	3.1	3.1	3.9
2-1/2	2.2	2.4	2.6	2.6	3.3	3.3	4.2
3	2.4	2.6	2.9	2.9	3.7	3.7	4.6
3-1/2	2.6	2.9	3.2	3.2	--	--	--
4	2.9	3.2	3.6	3.8	4.3	4.8	5.2
5	3.3	3.7	4.1	4.5	4.8	5.5	6.0
6	3.9	4.4	5.1	5.3	5.9	6.4	7.0
8	4.8	5.5	6.5	6.6	7.3	8.1	9.0
10	6.2	6.8	7.3	8.0	8.9	9.8	11.0
12	7.1	8.0	8.7	9.3	10.0	10.9	11.8
14 OD	8.5	9.6	10.1	11.7	13.1	14.4	--
16 OD	10.6	11.7	12.9	14.3	16.0	17.7	--
18 OD	12.3	13.3	14.5	15.9	17.3	19.1	--
20 OD	13.7	15.0	16.3	17.9	19.6	21.4	--
24 OD	19.1	20.4	21.5	22.7	24.8	26.6	--

Flanges: Man hours are for labor only. The price of the welding materials and flange must be added in all cases.

Pipe Thickness: Man hours are for any wall thickness of pipe used with listed flanges.

Unlisted Sizes: Unlisted sizes take the next higher listing.

ATTACHING FLANGES—SCREWED TYPE

Man Hours—Cutting and Threading Pipe, Making on Flange
Manual Seal Welding at Back and Refacing
Carbon Steel Material, Straight Pipe,
Bends, Headers and Nozzles

NET MAN HOURS EACH

Pipe Size Inches	SERVICE PRESSURE RATING						
	150 Lb.	300 Lb.	400 Lb.	600 Lb.	900 Lb.	1500 Lb.	2500 Lb.
2 or less	1.6	1.7	2.0	2.0	2.5	2.5	3.3
2-1/2	1.7	2.0	2.2	2.2	2.6	2.6	3.5
3	2.0	2.2	2.3	2.3	3.0	3.0	3.9
3-1/2	2.2	2.4	2.6	2.6	--	--	--
4	2.3	2.6	2.9	3.0	3.5	3.8	4.2
5	2.6	3.0	3.3	3.6	3.9	4.4	4.9
6	3.1	3.5	4.0	4.3	4.7	5.2	5.7
8	3.9	4.4	5.2	5.3	5.8	6.5	7.3
10	5.0	5.5	5.8	6.4	7.0	7.8	9.8
12	5.7	6.4	7.0	7.6	8.1	9.0	10.0
14 OD	6.8	7.7	8.5	9.6	10.5	12.6	--
16 OD	8.5	9.3	10.3	11.3	12.6	13.9	--
18 OD	9.9	10.6	11.4	12.5	13.8	15.5	--
20 OD	11.0	11.9	13.0	14.3	15.7	17.6	--
24 OD	15.3	16.3	17.2	18.4	19.8	21.6	--

Flanges: Man hours are for field labor only. The price of the flange must be added in all cases.

Pipe Thickness: Man hours are for any wall thickness of pipe used with listed flanges.

Unlisted Sizes: Unlisted sizes take the next higher listing.

ATTACHING FLANGES—SLIP-ON TYPE

Man Hours Slipping on Flange, Manual Welding at Front and Back

Carbon Steel Material, Straight Pipe, Bends, Headers and Nozzles

NET MAN HOURS EACH

Pipe Size Inches	SERVICE PRESSURE RATING						
	150 LB.	300 LB.	400 LB.	600 LB.	900 LB.	1500 LB.	2500 LB.
1	0.9	1.0	1.4	1.4	1.6	1.8	2.1
1-1/4	1.0	1.2	1.4	1.4	1.8	2.1	2.3
1-1/2	1.0	1.3	1.4	1.4	1.8	2.1	2.3
2	1.3	1.4	1.8	1.8	2.4	2.7	3.0
2-1/2	1.5	1.7	2.3	2.3	3.0	3.3	3.6
3	1.8	2.1	2.9	2.9	3.6	4.0	4.4
3-1/2	2.2	2.4	3.3	3.3	--	--	--
4	2.4	2.6	3.5	3.8	4.8	5.4	5.9
5	3.0	3.3	4.5	4.8	6.1	6.7	7.4
6	3.6	3.9	5.2	5.9	7.2	8.1	8.9
8	5.1	5.4	7.3	8.0	9.9	11.0	12.0
10	6.3	6.8	9.0	11.1	12.5	14.0	15.5
12	7.7	8.3	11.0	13.7	15.3	17.2	19.0
14 O.D.	9.0	10.0	13.0	16.2	17.7	19.8	--
16 O.D.	10.5	11.3	15.0	18.4	20.1	22.4	--
18 O.D.	12.2	13.5	17.5	21.1	23.7	26.6	--
20 O.D.	14.6	16.0	21.1	23.7	27.5	30.8	--
24 O.D.	18.3	20.1	25.6	31.2	34.8	38.9	--
26 O.D.	—	—	27.7	33.7	37.8	—	—
30 O.D.	—	—	32.0	38.9	43.5	—	—
34 O.D.	—	—	36.2	44.1	49.3	—	—
36 O.D.	—	—	38.4	46.7	52.2	—	—
42 O.D.	—	—	44.7	54.5	—	—	—

Flanges: Man hours are for field labor only. The price of welding materials and the flange must be added in all cases.

Pipe Thickness: Man hours are for any wall thickness of pipe used with listed flanges.

Preheating: If specified or required by codes, add for this operation. See man hours for pre-heating.

Stress Relieving: If specified or required by codes, add for this operation. See man hours for stress relieving.

Unlisted Sizes: Unlisted sizes take the next higher listing.

ATTACHING FLANGES—WELD NECK TYPE

Labor—Aligning Flange and Butt Welding

Carbon Steel Material

NET MAN HOURS EACH

Size Ins.	SERVICE PRESSURE RATING						
	150 Lb.	300 Lb.	400 Lb.	600 Lb.	900 Lb.	1500 Lb.	2500 Lb.
2	1.5	1.8	1.8	2.6	2.6	2.8	3.0
2½	2.0	2.3	2.3	3.4	3.4	3.6	4.2
3	2.5	2.8	2.8	4.1	4.1	4.3	4.5
4	3.2	3.5	3.5	5.0	5.0	5.6	5.8
6	4.2	4.7	4.7	6.7	6.7	7.4	7.6
8	5.4	6.0	6.0	8.6	8.6	9.8	10.2
10	6.7	7.3	7.3	10.1	10.1	11.6	11.8
12	7.3	7.9	7.9	10.5	10.5	12.3	13.3
14 OD	8.8	9.5	9.5	11.9	11.9	14.3	—
16 OD	9.6	10.2	10.2	12.3	12.3	16.1	—
18 OD	12.0	12.7	12.7	15.4	15.4	18.3	—
20 OD	13.3	14.0	14.0	16.9	16.9	21.6	—
24 OD	17.6	18.5	18.5	22.4	22.4	28.7	—
26 OD	—	—	20.5	23.2	23.2	—	—
30 OD	—	—	24.8	26.7	26.7	—	—
34 OD	—	—	30.8	32.8	32.8	—	—
36 OD	—	—	34.5	36.5	36.5	—	—
42 OD	—	—	49.8	52.1	—	—	—

Man hours include aligning, tack, and butt welding carbon steel weld neck flange to pipe.

Man hours are for any wall thickness of pipe used with listed flanges.

Unlisted sizes take the next highest listing.

ATTACHING ORIFICE FLANGES—SLIP-ON AND THREADED TYPES

Carbon Steel Material

MAN HOURS PER PAIR

Size Ins.	Slip-On Type	SERVICE PRESSURE RATING	Threaded Types	
	300 Lb.	300 Lb.	400–600 Lb.	900–1500 Lb.
1	3.8	5.2	—	7.2
1¼	4.2	5.2	—	7.2
1½	4.4	5.2	—	7.2
2	4.6	5.2	—	7.2
2½	5.3	5.8	—	7.7
3	7.1	6.3	—	11.0
4	8.6	8.6	9.8	12.7
6	12.0	10.7	12.4	16.0
8	16.2	13.2	17.2	19.7
10	20.6	17.0	21.0	24.2
12	24.4	21.4	25.2	28.8
14	29.3	24.7	—	—
16	33.5	28.3	—	—
18	38.2	32.6	—	—
20	45.2	37.0	—	—
24	54.6	47.0	—	—
26	71.2	—	—	—
30	80.7	—	—	—
34	92.0	—	—	—
36	98.8	—	—	—
42	107.6	—	—	—

Slip-On Types: Man hours include slipping on, welding, placement of paddle-type plates, and bolting of pair of orifice flanges.

Threaded Types: Man hours include screwing on, placement of paddle-type plates, and bolting up of pair of orifice flanges.

All man hours exclude cutting, beveling, or threading of pipe. See respective tables for these man hours.

ATTACHING ORIFICE FLANGES—WELD NECK TYPE

Carbon Steel Material
MAN HOURS PER PAIR

Size Ins.	SERVICE PRESSURE RATING				
	300 Lb.	400 Lb.	600 Lb.	900 Lb.	1500 Lb.
1	5.2	5.4	7.1	7.2	7.8
1¼	5.2	5.4	7.1	7.2	7.8
1½	5.2	5.4	7.1	7.2	7.8
2	5.2	5.4	7.1	7.2	7.8
3	7.5	8.0	10.2	10.4	11.1
4	10.4	10.9	12.5	12.7	14.3
6	13.1	13.8	16.2	16.5	18.4
8	17.4	17.9	20.8	21.2	24.3
10	19.6	20.0	25.9	26.9	28.8
12	22.4	25.0	27.1	28.7	32.4
14	25.6	27.3	30.4	32.1	37.1
16	28.3	29.3	36.8	37.6	41.6
18	34.8	35.8	39.6	41.5	47.8
20	38.2	39.2	44.4	45.4	55.5
24	49.4	50.4	57.7	58.9	—
26	—	54.8	59.8	60.8	—
30	—	63.3	67.8	70.1	—
34	—	76.2	80.9	83.4	—
36	—	80.0	89.8	91.6	—
42	—	116.8	122.3	—	—

Man hours include setting, aligning, welding, placement of paddle-type plates, and bolting up of pair of orifice flanges.

Man hours exclude cutting and beveling of pipe. See respective tables for these man hours.

GENERAL WELDING NOTES

Backing Rings: When backing rings are used, add 25% to the welding man hours to cover extra problems in fit-up. In addition the following percentages should be added if applicable.

    1) When backing rings are tack welded in on one side, add 10% to the man hours of a standard thickness butt weld.

    2) When backing rings are completely welded in on one side, add 30% to the man hours of a standard thickness butt weld.

    3) Preheating and stress relieving, when required, should be charged at full butt weld preheating and stress relieving man hours for the size and thickness in which the backing ring is installed.

Nozzle Welds: Following percentage increases should be allowed for the following conditions:

    1) When nozzle welds are to be located off-center of the run (except tangential) increase man hours shown for nozzle welds, 50%.

    2) Add 80% to nozzle welds for tangential nozzle welds.

    3) When nozzle welds are to be located on a fitting increase nozzle weld man hours 50%.

Long-neck Nozzle Welds: The welding-on of long neck nozzles should be charged at the schedule 160 reinforced nozzle weld man hours.

Shaped Nozzles, Nozzle Weld fit-ups and Dummy Nozzle Welds: These should be charged at a percentage of the completed nozzle weld man hours as follows:

    1) Shaped branch 50%
    2) Shaped hole in header 50%
    3) Fit-up of both branch or header (whether tack-welded or not) 60%
    4) Dummy nozzle weld (no holes in header) 70%

Sloping Lines: Add 100% to all welding man hours for this condition.

Consumable Inserts: When consumable inserts are used, add the following percentages to the welding man hours to cover extra problems in fit-up:

                1) Through 1/2" wall 40%
                2) Over 1/2" through 1" wall 30%
                3) Over 1" through 2" wall 20%
                4) Over 2" through 3" wall 15%
                5) Over 3" wall 10%

MANUAL BUTT WELDS

Man Hours—Welding Only
Carbon Steel Material

NET MAN HOURS EACH

Size Ins.	Standard Pipe & OD Sizes 3/8" Thick	Extra Heavy Pipe & OD Sizes 1/2" Thick	SCHEDULE NUMBERS								
			20	30	40	60	80	100	120	140	160
1	0.7	0.8	--	--	0.7	--	0.8	--	--	--	1.0
1-1/4	0.8	0.8	--	--	0.8	--	0.8	--	--	--	1.1
1-1/2	0.8	0.9	--	--	0.8	--	0.9	--	--	--	1.3
2	1.0	1.0	--	--	1.0	--	1.0	--	--	--	1.6
2-1/2	1.2	1.3	--	--	1.2	--	1.3	--	--	--	1.8
3	1.3	1.4	--	--	1.3	--	1.4	--	--	--	2.1
3-1/2	1.4	1.6	--	--	1.4	--	1.6	--	--	--	--
4	1.5	1.8	--	--	1.5	--	1.8	--	2.8	--	3.0
5	1.7	2.1	--	--	1.7	--	2.1	--	2.9	--	3.8
6	2.0	2.5	--	--	2.0	--	2.5	--	3.8	--	4.9
8	2.6	3.3	2.6	2.6	2.6	3.0	3.3	4.6	6.0	7.5	8.6
10	3.1	4.0	3.1	3.1	3.1	4.0	5.1	6.8	9.4	11.4	13.1
12	3.6	4.7	3.6	3.6	4.1	5.2	6.6	9.9	12.2	15.3	17.9
14 OD	4.3	5.7	4.3	4.3	5.0	6.8	9.6	13.2	16.2	19.2	22.7
16 OD	5.0	6.6	5.0	5.0	6.6	8.4	12.4	19.5	20.7	25.0	27.7
18 OD	5.9	7.7	5.9	6.8	8.6	11.2	16.4	21.8	25.6	29.9	33.7
20 OD	6.3	8.4	6.3	8.4	9.4	13.8	19.5	26.0	31.9	37.0	40.8
24 OD	6.9	10.1	6.9	--	13.3	20.1	25.2	35.8	43.5	49.3	59.3

Pipe Thickness: Wall thickness of the pipe determines the man hours that will apply, for butt welds of double extra strong materials, use schedule 160 man hours.

Mitre Welds: Add 50% to butt weld man hours.

Cutting and Beveling Pipe: Man hours do not include cutting and beveling of pipe. See respective tables for these charges.

Preheating: If specified or required by codes, add for this operation. See man hours for preheating.

Stress Relieving: Stress relieving of welds in carbon steel materials is required by the A.S.A. code for pressure piping, where the wall thickness is 3/4" or greater.

All sizes of butt welds shown below the ruled lines are 3/4" or greater in wall thickness and must be stress relieved.

Where stress relieving is required an extra charge should be made. See man hours for stress relieving.

For General Notes on welding, see page 92.

MANUAL HEAVY WALL BUTT WELDS

Labor for Welding Only

Carbon Steel Material

NET MAN HOURS EACH

Nominal Pipe Size	WALL THICKNESS IN INCHES							
	.750	1.00	1.25	1.50	1.75	2.00	2.25	2.50
3	2.7	3.7	--	--	--	--	--	--
4	3.3	4.1	5.7	6.8	--	--	--	--
5	--	4.7	6.7	8.0	10.0	12.4	--	--
6	--	6.4	8.5	10.4	13.3	15.6	18.2	--
8	--	8.7	10.1	13.1	16.5	19.2	22.7	27.4
10	--	--	13.5	16.2	20.1	23.2	27.3	32.1
12	--	--	--	19.6	23.2	27.4	32.6	37.5
14	--	--	--	23.5	26.6	31.2	36.5	43.1
16	--	--	--	--	29.9	35.6	41.5	49.7
18	--	--	--	--	--	39.8	46.4	54.8
20	--	--	--	--	--	46.4	54.8	66.2
22	--	--	--	--	--	--	--	72.3
24	--	--	--	--	--	--	--	78.7

	2.75	3.00	3.25	3.50	3.75	4.00	4.25	4.50
10	36.7	42.1	--	--	--	--	--	--
12	42.8	49.1	55.3	63.1	--	--	--	--
14	48.9	55.5	62.9	71.3	81.2	91.1	--	--
16	56.3	64.7	72.9	82.8	94.4	107.6	--	--
18	62.9	72.9	82.8	95.0	108.4	124.1	--	--
20	75.4	84.5	96.9	109.3	124.1	140.8	159.7	173.1
22	82.4	92.7	105.5	119.3	135.6	154.1	174.8	193.4
24	89.4	99.0	114.3	129.2	147.4	167.4	189.7	209.3

	4.75	5.00	5.25	5.50	5.75	6.00
20	189.7	203.3	216.9	225.9	251.5	268.1
22	204.8	219.8	240.6	252.6	276.8	295.1
24	223.5	240.6	262.9	275.5	298.2	319.0

For General Notes on welding, see pages 92 and 93.

MANUAL LARGE O.D. BUTT WELDS

Labor for Welding Only

Carbon Steel Material

NET MAN HOURS EACH

O.D. Pipe Inches	WALL THICKNESS IN INCHES							
	.375	.500	.750	1.00	1.25	1.50	1.75	2.00
26	8.4	11.4	15.1	20.2	26.7	34.5	43.4	52.5
28	10.0	13.1	16.4	22.2	29.3	37.3	46.4	55,7
30	12.5	15.2	18.9	24.1	31.7	39.8	49.6	58.9
32	15.5	17.9	21.5	26.7	34.9	43.0	52.7	62.1
34	19.4	21.5	24.4	29.5	39.3	46.1	56.3	65.4
36	23.0	24.7	27.8	33.2	45.2	52.0	62.3	71.7
38	27.0	28.9	32.0	37.1	52.0	58.8	68.6	78.1
40	31.6	34.2	36.8	41.5	59.7	66.3	75.4	85.2
42	36.9	40.4	42.5	46.6	68.8	75.0	82.8	92.9
44	42.8	46.6	49.9	57.0	74.9	83.2	90.3	101.2
46	48.3	53.1	58.3	67.9	82.7	91.5	98.3	109.6
48	54.5	59.9	68.1	79.1	90.9	99.9	106.8	118.1
54	61.4	67.6	79.5	92.2	99.7	109.3	116.0	127.3
60	69.0	76.2	92.9	107.4	109.5	119.4	126.0	137.2
	2.75	2.50	2.75	3.00	3.25	3.50	3.75	4.00
26	61.7	85.0	96.3	110.0	123.5	138.5	159.7	180.4
28	6 .0	91.0	104.4	117.5	138.2	150.3	174.8	195.2
30	68.3	99.5	112.9	126.5	144.2	161.1	185.3	209.3
32	71.4	104.0	118.9	132.9	153.0	170.2	196.9	222.9
34	76.9	110.0	126.5	142.2	161.4	180.4	209.3	237.9
36	83.0	117.5	134.9	150.7	171.7	192.2	222.5	250.0
	4.25	4.50	4.75	5.00	5.25	5.50	5.75	6.00
26	203.3	225.9	243.9	261.7	276.8	298.2	323.7	345.7
28	219.8	244.9	261.4	280.1	299.6	322.0	345.5	367.5
30	234.9	258.4	281.5	301.1	322.3	343.4	371.1	400.0
32	250.0	277.1	298.2	319.3	343.4	365.9	394.6	424.7
34	268.1	298.2	319.3	340.9	366.9	391.5	421.6	451.7
36	282.5	313.2	337.4	360.8	387.0	412.6	445.7	478.3

For General Notes on welding, see pages 92 and 93.

90° WELDED NOZZLES
Labor for Cutting and Welding Carbon Steel Material
NET MAN HOURS EACH

Size Ins.	Standard Pipe & OD Sizes 3/8" Thick	Extra Heavy Pipe & OD Sizes 1/2" Thick	SCHEDULE NUMBERS								
			20	30	40	60	80	100	120	140	160
1	2.1	2.2	--	--	2.1	--	2.2	--	--	--	3.1
1-1/4	2.2	2.4	--	--	2.2	--	2.4	--	--	--	3.6
1-1/2	2.4	2.6	--	--	2.4	--	2.6	--	--	--	4.0
2	2.5	3.1	--	--	2.5	--	3.1	--	--	--	5.3
2-1/2	2.8	3.8	--	--	2.8	--	3.8	--	--	--	5.9
3	3.2	4.4	--	--	3.2	--	4.4	--	--	--	6.6
3-1/2	3.7	4.9	--	--	3.7	--	4.9	--	--	--	--
4	4.0	5.6	--	--	4.0	--	5.6	--	7.0	--	8.6
5	5.1	6.9	--	--	5.1	--	6.9	--	8.6	--	10.7
6	5.4	7.5	--	--	5.4	--	7.5	--	10.9	--	13.9
8	6.3	8.9	6.3	6.3	6.3	8.3	8.9	12.0	15.2	18.6	21.5
10	7.1	10.3	7.1	7.1	7.1	10.3	12.6	16.4	21.1	27.3	32.8
12	8.1	11.8	8.1	8.1	9.9	13.1	17.0	23.5	28.7	34.7	39.2
14 OD	9.3	13.6	9.3	9.3	11.6	16.0	22.7	28.9	34.6	38.9	47.9
16 OD	10.6	15.2	10.6	10.6	15.2	20.2	26.8	36.5	41.4	45.7	55.3
18 OD	11.6	16.3	11.6	15.1	19.1	25.4	30.1	44.1	49.0	53.0	69.2
20 OD	13.0	18.3	13.0	18.3	22.3	32.6	35.4	51.0	55.9	60.8	77.7
24 OD	14.2	19.8	14.2	21.2	27.7	41.7	46.0	64.8	69.7	77.8	90.9

All nozzles other than 90° should be charged at the man hours shown for 45° nozzles.

Pipe Thickness: Wall thickness of the pipe used for the nozzle determines the man hours that will apply. For nozzles of double extra strong pipe thickness, use schedule 160 man hours.

Reinforcement: Man hours given above are for plain welded nozzles only. For use of gusset plates, etc., as stiffeners not for reinforcement, add 25% to the net man hours shown above. If reinforcement is required and produced by building up the nozzle weld, or by the use of reinforcing rings or saddles as specified use man hours for 90° reinforced nozzles.

Preheating: If specified or required by codes, add for this operation. See man hours for preheating. The size and wall thickness of the header (not the size of the nozzle) determines the preheating man hours.

Stress Relieving: Stress relieving of welds in carbon steel materials is required by the A.S.A. code for pressure piping, where the wall thickness is 3/4" or greater. The size and wall thickness of the header determines the man hours to be used for stress relieving.

All pipe sizes shown below the ruled line are 3/4" or greater in wall thickness and must be stress relieved. Where stress relieving is required an extra charge should be made. See man hours for stress relieving.

For General Notes on welding, see page 92.

90° WELDED NOZZLES—REINFORCED
Labor for Cutting and Welding Carbon Steel Material

NET MAN HOURS EACH

Size Ins.	Standard Pipe & OD Sizes 3/8" Thick	Extra Heavy Pipe & OD Sizes 1/2" Thick	SCHEDULE NUMBERS								
			20	30	40	60	80	100	120	140	160
1-1/2	5.0	5.4	--	--	5.0	--	5.4	--	--	--	7.7
2	5.3	5.9	--	--	5.3	--	5.9	--	--	--	9.9
2-1/2	5.8	7.0	--	--	5.8	--	7.0	--	--	--	10.8
3	6.7	7.9	--	--	6.7	--	7.9	--	--	--	12.0
3-1/2	7.5	8.7	--	--	7.5	--	8.7	--	--	--	--
4	8.1	10.0	--	--	8.1	--	10.0	--	12.5	--	15.3
5	9.8	11.8	--	--	9.8	--	11.8	--	15.1	--	18.4
6	10.2	12.8	--	--	10.2	--	12.8	--	18.5	--	23.1
8	11.7	14.5	11.7	11.7	11.7	13.4	14.5	19.8	24.5	29.3	32.8
10	12.7	16.3	12.7	12.7	12.7	16.3	20.2	26.0	31.0	34.8	39.1
12	14.2	18.3	14.2	14.2	15.3	20.3	26.3	35.5	39.5	44.6	47.7
14 OD	16.0	20.7	16.0	16.0	17.8	24.3	34.5	44.0	46.8	52.6	60.6
16 OD	17.9	23.2	17.9	17.9	22.8	30.1	40.1	54.8	57.5	60.6	70.0
18 OD	19.1	23.8	19.1	22.1	30.0	37.4	45.0	59.5	61.9	65.3	87.6
20 OD	21.1	26.6	21.1	26.4	32.3	47.2	55.5	63.8	73.8	84.6	98.4
24 OD	22.3	27.7	22.3	30.3	36.5	52.5	59.5	72.4	85.7	99.0	115.0

All Nozzles other than 90° should be charged at the man hours shown for 45° nozzles.

Pipe Thickness: Wall thickness of the pipe used for the nozzle determines the man hours that will apply. For nozzles of double extra strong pipe thickness use schedule 160 man hours.

Reinforcement: Man hours given above include the labor requirements for reinforcement produced by building up the nozzle weld, or by the use of reinforcing rings or saddles as may be specified.

Preheating: If specified or required by code, add for this operation. See man hours for preheating. The size and wall thickness of header (not the size of the nozzle) determines the preheating man hours.

Stress Relieving: Stress relieving of welds in carbon steel material is required by the A.S.A. code for pressure piping, where the wall thickness is 3/4" or greater. The size and wall thickness of the header determines the man hours to be used for stress relieving. All pipe sizes shown below the ruled line are 3/4" or greater in wall thickness and must be stress relieved. Where stress relieving is required an extra charge should be made. See man hours for stress relieving.

For General Notes on welding, see page 92.

LARGE O.D. 90° NOZZLE WELDS

Labor for Cutting and Welding

Carbon Steel Material

NET MAN HOURS EACH

NON-REINFORCED 90° NOZZLE WELDS

O.D. Pipe Inches	WALL THICKNESS IN INCHES								
	.375	.500	.750	1.00	1.25	1.50	1.75	2.00	2.25
26	24.8	29.0	31.9	45.4	53.3	68.4	78.7	89.0	104.5
28	28.1	33.2	34.7	50.0	58.6	74.7	84.1	94.3	110.1
30	33.2	38.2	40.0	54.2	63.5	79.4	89.8	99.7	115.8
32	38.7	43.3	45.9	59.8	69.7	86.0	95.6	104.9	120.7
34	46.6	48.1	51.7	66.4	78.6	92.4	102.1	110.6	130.2
36	53.1	55.1	59.1	74.7	90.3	103.8	113.0	121.3	140.4
38	60.8	62.2	68.0	84.4	104.0	117.4	125.6	132.3	151.7
40	68.7	70.3	78.2	95.3	119.5	132.6	139.4	144.2	163.9
42	77.6	79.4	89.9	107.7	137.5	149.9	154.7	157.1	177.0
48	85.0	90.6	102.5	123.0	156.9	171.1	176.8	179.6	202.3
54	95.6	102.0	115.3	138.3	176.5	192.5	198.8	202.0	227.5
60	106.2	113.3	128.1	153.6	196.1	213.8	220.9	224.4	252.8

REINFORCED 90° NOZZLE WELDS

	.375	.500	.750	1.00	1.25	1.50	1.75	2.00	2.25
26	34.3	40.4	44.1	53.6	62.5	80.8	102.0	123.3	145.1
28	38.8	46.0	48.1	58.5	68.4	87.6	108.9	131.0	152.7
30	45.9	53.0	55.5	63.6	74.6	93.5	116.5	138.4	160.6
32	53.6	59.8	63.1	70.6	82.0	100.9	123.9	145.8	167.7
34	64.8	66.9	71.7	78.0	92.3	108.6	132.5	153.4	180.9
36	74.6	76.5	81.8	87.6	106.1	122.1	146.2	168.4	194.8
38	81.1	87.2	93.3	99.0	122.0	138.1	161.0	183.6	210.4
40	95.3	99.4	106.4	111.9	140.4	156.0	177.0	200.1	227.3
42	107.6	113.3	121.3	126.4	161.4	176.3	194.8	218.2	245.6
48	123.0	129.7	138.8	144.4	184.7	201.7	222.5	249.2	280.4
54	138.3	146.0	156.1	162.5	207.7	226.8	250.4	280.4	315.4
60	153.6	162.1	173.5	180.5	230.8	252.0	278.2	311.5	350.5

For General Notes on welding, see pages 92, 96, and 97.

45° WELDED NOZZLES

Labor for Cutting and Welding

Carbon Steel Material

NET MAN HOURS EACH

Size Ins.	Standard Pipe & OD Sizes 3/8" Thick	Extra Heavy Pipe & OD Sizes 1/2" Thick	SCHEDULE NUMBERS								
			20	30	40	60	80	100	120	140	160
1	2.8	2.9	--	--	2.8	--	2.9	--	--	--	4.1
1-1/4	2.9	3.2	--	--	2.9	--	3.2	--	--	--	4.7
1-1/2	3.2	3.6	--	--	3.2	--	3.6	--	--	--	5.4
2	3.3	4.1	--	--	3.3	--	4.1	--	--	--	7.1
2-1/2	3.8	5.1	--	--	3.8	--	5.1	--	--	--	7.9
3	4.4	5.9	--	--	4.4	--	5.9	--	--	--	8.9
3-1/2	4.9	6.4	--	--	4.9	--	6.4	--	--	--	--
4	5.5	7.6	--	--	5.5	--	7.6	--	9.4	--	11.4
5	6.7	9.1	--	--	6.7	--	9.1	--	11.7	--	14.4
6	7.1	10.0	--	--	7.1	--	10.0	--	14.5	--	18.4
8	8.6	11.7	8.6	8.6	8.6	11.1	11.7	15.9	20.1	25.1	28.6
10	9.6	13.7	9.6	9.6	9.6	13.7	16.6	21.8	28.1	36.2	43.7
12	10.9	15.7	10.9	10.9	13.0	17.5	22.9	24.2	38.6	45.7	52.2
14 OD	12.4	18.2	12.4	12.4	15.3	21.4	30.1	38.4	45.9	51.6	64.2
16 OD	14.2	20.2	14.2	14.2	20.2	26.8	35.3	48.7	54.8	61.2	74.1
18 OD	15.5	21.6	15.5	20.1	25.5	34.3	40.6	59.1	65.7	71.0	92.7
20 OD	17.1	23.7	17.1	22.7	29.6	43.3	46.9	68.3	74.9	81.5	104.1
24 OD	18.8	26.6	18.8	31.8	35.5	48.0	55.0	86.8	93.4	104.3	121.8

Pipe Thickness: Wall thickness of the pipe used for the nozzle determines the man hours that will apply. For nozzles of double extra strong pipe thickness, use schedule 160 man hours.

Reinforcement: Man hours given above are for plain nozzles only. For use of gusset plates, etc., as stiffeners, not for reinforcement, add 25% to the net man hours shown above. If reinforcement is required and produced by building up the nozzle weld, or by the use of reinforcing rings or saddles as specified use man hours for 45° reinforced nozzles.

Preheating: If specified or required by codes, add for this operation. See man hours for preheating. The size and wall thickness of the header (not the size of the nozzle) determines the preheating man hours.

Stress Relieving: Stress relieving of welds in carbon steel material is required by the A.S.A. code for pressure piping, where the wall thickness is 3/4" or greater. The size and wall thickness of the header determines the man hours to be used for stress relieving. All pipe sizes shown below the ruled line are 3/4" or greater in wall thickness and must be stress relieved. Where stress relieving is required an extra charge should be made. See man hours for stress relieving.

General Notes: For additional notes on welding see page 92.

45° WELDED NOZZLES—REINFORCED

Labor for Cutting and Welding

Carbon Steel Material

NET MAN HOURS EACH

Size Ins.	Standard Pipe & OD Sizes 3/8" Thick	Extra Heavy Pipe & OD Sizes 1/2" Thick	SCHEDULE NUMBERS								
			20	30	40	60	80	100	120	140	160
1-1/2	6.8	7.1	--	--	6.8	--	7.1	--	--	--	10.2
2	6.9	7.7	--	--	6.9	--	7.7	--	--	--	13.3
2-1/2	7.8	9.3	--	--	7.8	--	9.3	--	--	--	14.6
3	8.7	10.5	--	--	8.7	--	10.5	--	--	--	16.1
3-1/2	10.0	11.6	--	--	10.0	--	11.6	--	--	--	--
4	11.2	13.5	--	--	11.2	--	13.5	--	--	--	20.2
5	13.1	15.8	--	--	13.1	--	15.8	--	--	--	25.0
6	13.8	17.1	--	--	13.8	--	17.1	--	--	--	31.4
8	15.8	21.6	15.8	15.8	15.8	18.1	21.6	26.4	32.8	39.8	44.5
10	17.3	21.7	17.3	17.3	17.3	21.7	26.0	32.1	44.5	48.9	53.2
12	19.2	24.3	19.2	19.2	20.5	27.1	35.4	45.7	51.7	60.8	70.9
14 OD	21.0	27.6	21.0	21.0	23.4	32.6	45.9	52.6	62.2	70.5	81.2
16 OD	23.7	30.2	23.7	23.7	30.2	40.1	52.9	65.7	82.6	84.8	93.8
18 OD	25.5	31.6	25.5	29.4	37.4	50.2	59.5	79.7	82.9	87.5	117.4
20 OD	28.2	35.6	28.2	35.6	43.4	62.9	71.3	85.5	98.9	113.4	131.9
24 OD	29.5	38.4	29.5	38.7	51.1	71.2	76.8	97.1	114.8	132.7	154.1

Pipe Thickness: Wall thickness of the pipe used for the nozzle determines the man hours that will apply. For nozzles of double extra strong pipe thickness, use schedule 160 man hours.

Reinforcement: Man hours given above includes the labor requirements for reinforcement produced by building up the nozzle weld, or by the use of reinforcing rings or saddles as may be specified.

Preheating: If specified or required by codes, add for this operation. See man hours for preheating. The size and wall thickness of the header (not the size of the nozzles) determines the preheating man hours.

Stress Relieving: Stress relieving of welds in carbon steel material is required by the A. S. A. code for pressure piping, where the wall thickness is 3/4" or greater. The size and wall thickness of the header determines the man hours to be used for stress relieving. All pipe sizes shown below the ruled line are 3/4" or greater in wall thickness and must be stress relieved. Where stress relieving is required an extra charge should be made. See man hours for stress relieving.

For General Notes on welding, see page 92.

LARGE O.D. 45° NOZZLE WELDS

Labor for Cutting and Welding

Carbon Steel Material

NET MAN HOURS EACH

NON-REINFORCED 45° NOZZLE WELDS

O.D. Pipe Inches	WALL THICKNESS IN INCHES								
	.375	.500	.750	1.00	1.25	1.50	1.75	2.00	2.25
26	33.2	38.7	42.6	60.8	71.0	91.8	105.1	118.6	139.5
28	37.2	44.4	46.1	66.6	77.6	99.4	112.2	125.7	146.8
30	44.1	51.0	53.2	72.2	84.6	106.2	119.9	133.0	154.2
32	51.4	57.6	60.7	80.1	93.0	114.7	127.4	140.2	161.1
34	62.2	64.4	68.8	88.5	104.8	123.1	136.3	147.4	173.8
36	71.6	73.4	78.7	99.4	120.4	138.5	150.5	161.9	187.0
38	80.9	83.8	90.6	112.3	138.5	156.6	165.7	176.5	202.3
40	91.5	95.5	104.2	127.0	159.4	176.9	182.2	192.3	218.3
42	103.4	108.8	119.9	143.4	183.3	199.9	207.8	224.7	235.8
48	118.4	124.6	137.1	163.7	209.6	226.6	237.3	256.5	269.0
54	133.2	140.2	154.2	184.2	235.8	254.9	267.0	288.6	302.7
60	148.0	155.8	171.3	204.6	262.0	283.2	296.7	320.7	336.3

REINFORCED 45° NOZZLE WELDS

O.D. Pipe Inches	.375	.500	.750	1.00	1.25	1.50	1.75	2.00	2.25
26	45.9	53.7	58.9	75.0	83.3	107.6	136.1	164.4	193.3
28	51.6	61.4	64.1	78.0	91.3	116.8	145.3	174.5	203.8
30	61.1	70.7	74.0	84.8	99.4	124.8	155.2	184.4	214.1
32	71.5	79.8	84.3	94.0	109.4	134.5	165.2	194.5	223.5
34	86.3	89.3	95.6	104.0	123.1	144.7	176.6	204.6	241.1
36	99.2	101.8	109.2	116.8	141.5	162.7	194.9	224.4	259.8
38	112.1	116.2	124.5	132.0	162.7	183.8	214.6	244.7	280.7
40	126.7	132.4	142.0	149.2	187.1	207.8	236.0	266.7	303.1
42	143.3	150.9	161.9	168.6	215.2	234.8	259.6	290.8	327.5
48	163.7	172.8	185.3	192.6	245.8	268.5	296.8	332.5	374.4
54	184.2	194.3	208.4	216.6	276.6	302.1	333.9	374.1	421.1
60	204.6	215.9	231.5	240.7	307.3	335.6	371.0	415.6	468.0

For General Notes on welding, see pages 92, 96, and 97.

CONCENTRIC SWEDGED ENDS

Labor For Welding

Carbon Steel Material

NET MAN HOURS EACH

Size Ins.	Standard Pipe & OD Sizes 3/8" Thick	Extra Heavy Pipe & OD Sizes 1/2" Thick	SCHEDULE NUMBERS								
			20	30	40	60	80	100	120	140	160
2	1.6	2.1	--	--	1.6	--	2.1	--	--	--	3.2
2-1/2	1.8	2.4	--	--	1.8	--	2.4	--	--	--	3.8
3	2.0	2.8	--	--	2.0	--	2.8	--	--	--	4.5
3-1/2	2.3	3.2	--	--	2.3	--	3.2	--	--	--	--
4	2.6	3.8	--	--	2.6	--	3.8	--	5.4	--	6.4
5	3.5	4.8	--	--	3.5	--	4.8	--	7.6	--	9.0
6	4.1	6.2	--	--	4.1	--	6.2	--	10.2	--	11.6
8	5.9	9.2	--	5.9	5.9	--	9.2	11.9	15.6	18.3	20.1
10	7.8	12.4	--	7.8	7.8	12.4	14.6	18.6	26.7	--	34.8
12	10.3	16.5	--	10.3	15.6	19.4	24.8	32.9	43.9	--	50.4
14 OD	13.6	22.9	13.6	13.6	22.1	27.5	35.8	45.0	62.3	--	--
16 OD	19.4	29.3	19.4	19.4	29.3	34.7	39.5	51.2	68.8	--	--
18 OD	23.7	38.5	23.7	35.8	45.0	64.1	--	--	--	--	--
20 OD	27.5	43.1	27.5	43.1	50.4	76.8	--	--	--	--	--
24 OD	36.6	59.6	36.6	59.6	--	--	--	--	--	--	--

Pipe Thickness: The wall thickness of the pipe determines the man hours that will apply. For swedged ends on double extra strong pipe thickness, use schedule 160 man hours.

Ends: All man hours are based on ends being furnished either plain or beveled for welding.

Unlisted Sizes: Unlisted sizes take the next higher listing.

ECCENTRIC SWEDGED ENDS

Labor For Welding

Carbon Steel Material

NET MAN HOURS EACH

Size Ins.	Standard Pipe & OD Sizes 3/8" Thick	Extra Heavy Pipe & OD Sizes 1/2" Thick	SCHEDULE NUMBERS								
			20	30	40	60	80	100	120	140	160
2	1.8	2.2	--	--	1.8	--	2.2	--	--	--	3.9
2-1/2	2.0	2.9	--	--	2.0	--	2.9	--	--	--	4.6
3	2.3	3.2	--	--	2.3	--	3.2	--	--	--	5.2
3-1/2	2.6	3.8	--	--	2.6	--	3.8	--	--	--	--
4	3.1	4.5	--	--	3.1	--	4.5	--	7.1	--	7.6
5	4.0	6.0	--	--	4.0	--	6.0	--	9.4	--	10.4
6	4.9	7.1	--	--	4.9	--	7.1	--	12.8	--	14.4
8	7.3	11.9	7.3	7.3	7.3	--	11.9	20.5	20.1	22.1	24.1
10	10.0	16.2	10.0	10.0	10.0	16.2	20.1	24.8	32.1	--	45.8
12	13.7	21.1	13.7	13.7	21.1	25.6	32.9	43.9	56.8	--	64.0
14 OD	19.4	30.2	19.4	19.4	31.0	37.5	45.8	58.6	80.5	--	--
16 OD	27.4	40.2	27.4	27.4	40.2	47.6	52.3	68.8	89.7	--	--
18 OD	32.5	54.9	32.5	32.5	60.4	85.2	--	--	--	--	--
20 OD	36.0	62.9	36.0	36.0	68.8	100.7	--	--	--	--	--
24 OD	51.3	82.4	51.3	51.3	--	--	--	--	--	--	--

Pipe Thickness: The wall thickness of the pipe determines the man hours that will apply. For swedged ends on double extra strong pipe thickness, use schedule 160 man hours.

Ends: All man hours are based on ends being furnished either plain or beveled for welding.

Unlisted Sizes: Unlisted sizes take the next higher listing.

END CLOSURES—PRESSURE TYPE

Carbon Steel Material

NET MAN HOURS EACH

Nom. Pipe Size Ins.	Standard Pipe & OD Sizes 3/8" Thick	Extra Heavy Pipe & OD Sizes 1/2" Thick	SCHEDULE NUMBERS						
			40	60	80	100	120	140	XX Hy. or 160
1-1/2	0.9	1.0	0.9	--	0.9	--	--	--	2.3
2	1.2	1.3	1.2	--	1.3	--	--	--	3.5
2-1/2	1.3	1.5	1.3	--	1.5	--	--	--	4.3
3	1.5	1.8	1.5	--	1.8	--	--	--	4.5
3-1/2	1.6	2.1	1.6	--	2.1	--	--	--	--
4	1.8	2.4	1.8	--	2.4	--	5.6	--	6.0
5	2.3	2.9	2.3	--	2.9	--	7.5	--	7.9
6	2.6	3.3	2.6	--	3.3	--	9.3	--	10.2
8	3.7	4.7	3.7	--	4.7	8.9	12.5	14.8	15.6
10	4.6	5.9	4.6	5.9	10.3	12.7	21.4	23.0	24.7
12	5.5	7.2	6.7	7.7	13.0	18.9	28.6	31.2	33.7
14	6.6	8.6	8.0	9.9	15.1	20.9	32.9	44.5	46.3
16	7.4	9.7	9.7	12.5	16.6	23.5	37.9	56.1	58.8
18	9.0	11.3	13.9	16.2	21.8	29.6	42.8	67.6	71.3
20	9.7	12.5	16.8	19.8	27.0	35.8	47.8	79.1	--
24	10.6	14.9	19.6	23.5	32.2	42.1	52.2	--	--

Pipe Thickness: Wall thickness of pipe determines the man hours that will apply. For double strong pipe thickness use schedule 160 man hours.

Construction: End closures such as orange peel, saddle, or flat plate type.

Preheating: If specified or required by codes, add for this operation. See man hours for preheating.

Stress Relieving: Stress relieving of welds in carbon steel material is required by the A. S. A. Code for pressure piping where the wall thickness is 3/4" or greater.

All sizes of butt welds shown below the ruled lines are 3/4" or greater in wall thickness and must be stress relieved, if the end closure involves a circumferential weld. Where stress relieving is required, an extra charge should be made. See man hours for stress relieving.

Unlisted Sizes: Unlisted sizes take the next higher listing.

HEAVY WALL END CLOSURES—PRESSURE TYPE

Carbon Steel Material
NET MAN HOURS EACH

Nominal Pipe Size	WALL THICKNESS IN.							
	.750	1.00	1.25	1.50	1.75	2.00	2.25	2.50
3	5.4	—	—	—	—	—	—	—
4	—	8.1	9.6	10.9	—	—	—	—
5	—	9.9	11.6	13.1	14.8	—	—	—
6	—	15.8	18.5	21.0	23.7	25.8	27.4	—
8	—	16.6	19.5	22.1	24.9	27.1	28.8	30.8
10	—	—	26.3	29.9	33.7	36.8	39.1	41.8
12	—	—	—	41.3	46.7	51.0	54.0	57.8
14	—	—	—	47.7	53.9	58.8	62.3	66.7
16	—	—	—	—	62.8	68.4	72.6	77.5
18	—	—	—	—	—	76.2	80.8	86.4
20	—	—	—	—	—	82.2	87.2	93.2
22	—	—	—	—	—	90.5	95.9	102.5
24	—	—	—	—	—	98.8	104.7	111.9
	2.75	3.00	3.25	3.50	3.75	4.00	4.25	4.50
10	44.6	47.7	—	—	—	—	—	—
12	61.8	66.1	69.1	72.2	—	—	—	—
14	62.3	66.6	69.7	72.9	76.2	79.8	—	—
16	82.8	88.5	92.6	96.8	101.1	105.7	—	—
18	92.4	98.8	103.4	108.0	112.8	117.9	—	—
20	99.7	106.6	111.5	116.6	121.8	127.2	132.3	136.9
22	109.6	117.2	122.7	128.3	134.0	140.1	145.6	150.7
24	119.5	127.8	133.8	139.8	146.1	152.7	158.8	164.4
	4.75	5.00	5.25	5.50	5.75	6.00		
20	142.5	147.5	152.7	157.3	162.0	166.9		
22	156.8	162.4	168.0	173.1	178.3	183.6		
24	171.1	177.1	183.4	188.9	194.6	200.4		

Construction: End closures as such are field fabricated closures; orange peel, saddle, or flat plate type.

Preheating: If specified or required by codes, add for this operation. See man hours for preheating.

Stress Relieving: Stress relieving of welds in carbon steel material is required by the A.S.A. Code of Pressure Piping where the wall thickness is ¾" or greater.

All the above butt welds are ¾" or greater and must be stress relieved, if end closure involves a circumferential weld.

See respective man hour tables for stress relieving.

LARGE O.D. PIPE END CLOSURES—PRESSURE TYPE

Carbon Steel Material
NET MAN HOURS EACH

O.D. Pipe In.	WALL THICKNESS IN.							
	.375	.500	.750	1.00	1.25	1.50	1.75	2.00
26	33.3	40.7	53.2	65.0	76.0	86.6	96.2	105.6
28	35.2	43.0	57.0	69.7	81.5	92.6	102.8	113.4
30	36.9	45.2	60.1	73.5	85.9	97.6	108.2	119.3
32	38.5	47.1	62.5	76.5	89.4	101.6	112.7	124.3
34	40.0	49.0	65.0	79.5	93.0	105.6	117.2	129.2
36	41.4	50.6	67.3	82.2	96.2	109.3	121.2	133.7
38	42.8	52.4	69.6	85.2	100.0	113.2	125.6	138.5
40	44.6	54.5	72.5	88.6	103.6	117.6	130.5	144.0
42	46.4	56.8	75.4	92.2	107.7	122.4	135.7	149.6
44	47.9	58.6	77.9	95.2	111.3	126.4	140.2	154.6
46	87.4	60.7	80.6	98.5	115.2	130.9	145.1	160.1
48	51.2	62.7	83.2	101.7	118.9	135.1	149.9	165.3
54	54.8	66.9	88.9	108.7	127.1	144.3	160.0	176.5
60	58.5	71.6	95.1	116.3	136.1	154.6	171.5	189.2
	2.25	2.50	2.75	3.00	3.25	3.50	3.75	4.00
26	114.7	124.8	133.5	142.0	151.9	159.2	166.9	174.9
28	123.2	134.0	143.3	152.5	163.1	170.9	179.1	187.7
30	129.6	141.0	150.7	160.4	171.6	179.8	188.4	197.5
32	135.0	146.9	157.1	167.1	178.8	187.4	196.4	205.8
34	140.3	152.7	163.2	173.7	185.9	194.8	204.1	213.9
36	145.1	157.9	168.7	179.6	192.2	200.7	210.4	220.5
	4.25	4.50	4.75	5.00	5.25	5.50	5.75	6.00
26	182.1	189.5	197.3	204.0	211.0	218.2	225.6	232.0
28	195.4	203.4	211.8	219.0	226.4	234.1	242.0	248.7
30	205.7	214.1	222.8	230.3	238.1	246.3	254.6	261.7
32	214.3	223.0	232.1	240.0	248.2	256.7	265.4	272.8
34	222.7	231.8	241.3	249.6	258.1	266.8	275.9	283.6
36	229.6	239.1	248.9	257.4	266.1	275.2	284.5	292.5

Construction: End closures as such are field fabricated closures; orange peel, saddle, or flat plate type.

Preheating: If specified or required by codes, add for this operation. See man hours for preheating.

Stress Relieving: Stress relieving of welds in carbon steel material is required by the A.S.A. Code of Pressure Piping where the wall thickness is ¾″ or greater.

Above wall thickness .750 through 6.00 must be stress relieved, if the end closure involves a circumferential weld.

See respective table for stress relieving.

90° COUPLING WELDS AND SOCKET WELDS

Labor for Cutting and Welding

Carbon Steel Material

NET MAN HOURS EACH

Pipe Size Inches	90°—3000 # Coupling Weld	90°—6000 # Coupling Weld	SOCKET WELDS	
			Sch. 40 & 80 Pipe	Sch. 100 & Heavier Pipe
1/2" or less	1.6	2.0	0.6	0.6
3/4	1.8	2.2	0.6	0.7
1	2.1	2.5	0.7	0.8
1-1/4	2.4	2.9	0.9	1.0
1-1/2	2.6	3.2	0.9	1.2
2	3.3	4.1	1.0	1.5
2-1/2	3.9	4.8	1.3	1.6
3	4.6	5.6	1.4	2.0

Man hours shown are for welding of coupling to the O.D. of the pipe only.

If couplings are to be welded to the I.D. of the pipe, add 50% to the above man hours. For pipe thickness up to 1 inch, add an additional 12% for each 1/4 inch or fraction thereof of pipe thickness over 1 inch.

Any coupling welded to pipe heavier than schedule 160 should be man houred as a 6000 pound coupling.

For couplings welded at angles from 45° to less than 90° and couplings attached to fittings increase above man hours 50%.

For couplings welded at angles less than 45% increase above man hours 75%.

Socket welds do not include cut. See respective man hour table for this charge.

'OLET TYPE WELDS

Labor Cutting And Welding

Carbon Steel Material

NET MAN HOURS EACH

Nominal Pipe Size		Standard Weight And 2000 #	Extra Strong and 3000 #	Greater Than Extra Strong and 6000 #
Outlet	Header			
1/2	All Sizes	1.5	2.0	2.5
3/4	All Sizes	1.8	2.2	3.0
1	All Sizes	2.1	2.5	3.3
1-1/4	All Sizes	2.3	2.9	3.8
1-1/2	All Sizes	2.9	3.7	4.9
2	All Sizes	3.9	4.8	6.4
2-1/2	All Sizes	4.6	5.9	7.7
3	All Sizes	5.3	6.8	10.6
4	All Sizes	7.0	8.5	11.3
5	All Sizes	7.9	9.3	13.7
6	All Sizes	8.7	9.9	16.0
8	All Sizes	9.7	10.6	18.9
10	All Sizes	13.6	19.4	30.2
12	All Sizes	19.0	22.5	44.7
14	14" And 16"	23.8	26.5	53.9
14	18" And Larger	21.2	23.8	58.7
16	16" And 18"	28.4	30.4	70.4
16	20" And Larger	25.1	27.4	76.2
18	18" And 20"	33.7	36.9	91.0
18	24" And Larger	29.7	32.7	98.0
20	20" And 24"	40.9	44.9	100.9
20	26" And Larger	35.7	39.9	108.8
24	24" And 26"	62.7	73.3	121.1
24	28" And Larger	52.8	63.4	130.5

Man hours are based on the outlet size and schedule except when the run schedule is greater than the outlet schedule, in which case the man hours are based on the outlet size and run schedule.

For elbolet or latrolet welds and weldolets, thredolets, etc., that are attached to fittings or welded at any angle other than 90°, add 50% to the above man hours.

For sweepolet attachment welds, add 150% to the above man hours.

FLAME CUTTING PIPE—PLAIN ENDS

Labor For Straight Pipe Only

Carbon Steel Material

NET MAN HOURS EACH

Pipe Size Inches	Standard Pipe & O.D. Size 3/8" Thick	Extra Hvy. Pipe & O.D. Sizes 1/2" Thick	SCHEDULE NUMBERS								
			20	30	40	60	80	100	120	140	160
2" Or Less	0.10	0.15	--	--	0.10	--	0.15	--	--	--	0.21
2-1/2	0.12	0.17	--	--	0.12	--	0.17	--	--	--	0.23
3	0.15	0.21	--	--	0.15	--	0.21	--	--	--	0.28
4	0.21	0.28	--	--	0.21	--	0.28	--	0.38	--	0.41
5	0.24	0.36	--	--	0.24	--	0.36	--	0.44	--	0.49
6	0.33	0.45	--	--	0.33	--	0.45	--	0.56	--	0.63
8	0.46	0.64	0.46	0.46	0.46	0.59	0.64	0.76	0.86	0.97	1.14
10	0.64	0.92	0.64	0.64	0.64	0.92	0.99	1.09	1.24	1.43	1.73
12	0.70	1.09	0.70	0.70	0.86	1.30	1.37	1.48	1.73	1.91	2.05
14 O.D.	0.98	1.30	0.98	0.98	1.15	1.44	1.67	1.78	1.96	2.30	2.42
16 O.D.	1.09	1.61	1.09	1.09	1.61	1.78	1.90	2.13	2.30	2.59	2.93
18 O.D.	1.42	2.01	1.42	1.65	2.01	2.24	2.36	2.66	2.83	3.19	3.72
20 O.D.	1.71	2.24	1.71	2.30	2.48	2.66	2.83	3.13	3.30	3.84	4.37
24 O.D.	2.60	3.30	2.60	3.48	3.66	3.84	3.95	4.31	4.78	5.37	6.08

For mitre cuts less than 30°, add 50% to the above man hours.

For mitre cuts 30° or greater, add 100% to the above man hours.

Man hours are for cutting pipe with plain ends only and do not include beveling, threading, etc. See appropriate man hour tables for these operations.

For cutting the ends of bends or trimming fittings, add 50% to the above man hours.

FLAME CUTTING HEAVY WALL PIPE—PLAIN ENDS

Labor For Straight Pipe Only

Carbon Steel Material

NET MAN HOURS EACH

Nominal Pipe Size	WALL THICKNESS IN INCHES							
	.750	1.00	1.25	1.50	1.75	2.00	2.25	2.50
3	0.53	0.94	--	--	--	--	--	--
4	0.94	1.12	1.59	1.77	--	--	--	--
5	--	1.30	1.65	2.00	2.24	2.54	--	--
6	--	1.59	1.89	2.24	2.54	2.77	3.19	--
8	--	1.89	2.24	2.48	2.89	3.13	3.48	4.07
10	--	--	2.48	2.77	3.13	3.42	3.84	4.37
12	--	--	--	3.13	3.36	3.84	4.25	4.78
14	--	--	--	3.60	3.72	4.13	4.60	5.25
16	--	--	--	--	4.31	4.96	5.61	6.14
18	--	--	--	--	--	4.60	6.14	6.84
20	--	--	--	--	--	6.14	6.84	7.73
22	--	--	--	--	--	--	--	8.38
24	--	--	--	--	--	--	--	9.50
	2.75	3.00	3.25	3.50	3.75	4.00	4.25	4.50
10	4.72	5.02	--	--	--	--	--	--
12	5.31	5.72	6.14	6.61	--	--	--	--
14	5.84	6.31	6.61	7.49	8.02	8.79	--	--
16	6.67	7.20	7.73	8.38	9.09	9.97	--	--
18	7.49	8.02	8.79	9.50	10.27	11.39	--	--
20	8.26	8.91	9.79	10.68	11.56	12.57	13.63	14.63
22	9.26	10.03	10.68	11.86	12.92	13.92	14.99	16.11
24	10.27	11.15	12.10	13.22	13.92	15.46	16.87	18.47
	4.75	5.00	5.25	5.50	5.75	6.00		
20	15.64	16.58	17.70	18.64	19.77	20.77		
22	17.23	18.35	19.35	20.41	21.54	22.72		
24	19.94	21.00	22.30	23.60	24.78	26.14		

For mitre cuts less than 30°, add 50% to the above man hours.

For mitre cuts 30° or greater, add 100% to the above man hours.

Man hours are for cutting pipe with plain ends only and do not include beveling, threading, etc. See appropriate man hour tables for these operations.

For cutting the ends of bends or trimming fittings, add 50% to the above man hours.

FLAME CUTTING LARGE O.D. PIPE-PLAIN ENDS

Labor for Straight Pipe Only

Carbon Steel Material

NET MAN HOURS EACH

O.D. Pipe Inches	WALL THICKNESS IN INCHES							
	.375	.500	.750	1.00	1.25	1.50	1.75	2.00
26	4.13	5.49	6.20	6.67	6.84	7.20	7.49	7.73
28	4.78	5.96	6.67	7.20	7.49	7.29	7.91	8.20
30	5.07	6.43	7.20	7.49	7.91	8.20	8.44	8.79
32	5.61	6.84	7.55	8.02	8.38	8.79	9.09	9.26
34	6.14	7.55	8.14	8.61	8.91	9.26	9.62	9.85
36	6.84	8.20	8.79	9.26	9.62	9.97	10.27	10.62
38	7.72	8.44	9.50	9.97	10.27	10.74	11.33	11.68
40	8.61	9.62	10.38	10.68	11.15	11.62	12.10	12.57
42	9.62	10.97	11.39	11.68	12.10	12.74	13.22	13.63
44	10.92	12.04	12.57	12.92	13.45	13.87	14.34	14.81
46	12.21	13.22	13.81	14.10	14.57	15.16	15.69	16.11
48	13.81	14.51	14.99	15.46	15.93	16.40	17.05	17.46
54	15.53	16.32	16.85	17.39	17.91	18.44	19.18	19.65
60	17.25	18.14	18.73	19.32	19.91	20.50	21.31	21.83

	2.25	2.50	2.75	3.00	3.25	3.50	3.75	4.00
26	8.14	10.50	11.39	12.27	13.22	14.34	14.99	16.58
28	8.79	11.15	11.63	12.57	13.57	14.81	15.52	16.99
30	9.26	11.68	12.04	12.92	14.16	15.28	15.93	17.52
32	9.79	12.27	12.74	13.45	14.63	15.75	16.40	18.00
34	10.38	12.92	13.39	13.87	15.22	16.23	16.87	18.47
36	10.97	13.45	13.92	14.34	15.69	16.87	17.52	19.12

	4.25	4.50	4.75	5.00	5.25	5.50	5.75	6.00
26	18.00	19.53	20.95	22.30	23.72	25.19	26.61	28.03
28	18.47	20.06	21.42	22.95	24.19	25.67	27.08	28.50
30	18.94	20.41	22.00	23.42	24.78	26.14	27.55	28.97
32	19.53	21.00	22.48	23.90	25.25	26.67	28.03	29.50
34	20.06	21.59	23.00	24.37	25.78	27.26	28.67	30.09
36	20.53	22.13	23.60	25.00	26.43	27.73	29.21	30.74

For General Notes, see the bottom of page 110.

FLAME BEVELING PIPE FOR WELDING

"V" Type Bevels

Labor for Straight Pipe Only

Carbon Steel Material

NET MAN HOURS EACH

Pipe Size Inches	Standard Pipe & O.D. Sizes 3/8" Thick	Extra Hvy. Pipe & O.D. Sizes 1/2" Thick	SCHEDULE NUMBERS								
			20	30	40	60	80	100	120	140	160
2" or less	0.08	0.12	--	--	0.08	--	0.12	--	--	--	0.16
2-1/2"	0.09	0.14	--	--	0.09	--	0.14	--	--	--	0.18
3	0.12	0.16	--	--	0.12	--	0.16	--	--	--	0.22
4	0.16	0.22	--	--	0.16	--	0.22	--	0.30	--	0.32
5	0.20	0.28	--	--	0.20	--	0.28	--	0.35	--	0.39
6	0.26	0.35	--	--	0.26	--	0.35	--	0.45	--	0.49
8	0.37	0.51	0.37	0.37	0.37	0.46	0.51	0.60	0.68	0.75	--
10	0.51	0.72	0.51	0.51	0.51	0.72	0.78	0.86	0.95	--	--
12	0.55	0.86	0.55	0.55	0.68	1.02	1.08	1.18	--	--	--
14 O.D.	0.57	1.02	0.77	0.77	0.91	1.13	1.31	--	--	--	--
16 O.D.	0.86	1.27	0.86	0.86	1.27	1.40	1.55	--	--	--	--
18 O.D.	1.11	1.58	1.11	1.30	1.58	1.77	--	--	--	--	--
20 O.D.	1.35	1.77	1.35	1.82	1.95	2.15	--	--	--	--	--
24 O.D.	2.04	2.60	2.04	2.74	2.88	--	--	--	--	--	--

For mitre bevels add 50% to the above man hours.

Above man hours are for flame "V" beveling only and do not include cutting or internal machining. See respective man hour tables for these charges.

For beveling on the ends of bends or shop trimmed fittings, add 50% to the above man hours.

FLAME BEVELING LARGE O.D. PIPE FOR WELDING

Labor For Straight Pipe Only

Carbon Steel Material

NET MAN HOURS EACH

O. D. Pipe Size Inches	WALL THICKNESS IN INCHES		
	.375	.500	.750
26	3.27	4.31	4.87
28	3.75	4.68	5.24
30	4.00	5.05	5.69
32	4.43	5.39	5.95
34	4.84	5.95	6.43
36	5.39	6.47	6.91
38	6.10	7.17	7.47
40	6.80	7.85	8.18
42	7.59	8.66	8.96
44	8.59	9.48	9.89
46	9.63	10.41	10.89
48	10.89	11.45	11.82
54	12.25	12.87	13.31
60	13.62	14.31	14.79

For Mitre Bevels add 50% to the above man hours.

Above man hours are for flame "V" beveling only and do not include cutting or internal machining. See respective man hour tables for these charges.

For beveling on the ends of bends or shop trimmed fittings, add 50% to the above man hours.

THREADING PIPE—INCLUDING CUT

Labor For Cut and Thread Only

Carbon Steel Material

NET MAN HOURS EACH

Pipe Size Inches	Standard Pipe & O.D. Sizes 3/8" Thick	Extra Hvy. Pipe & O.D. Sizes 1/2" Thick	SCHEDULE NUMBERS								
			20	30	40	60	80	100	120	140	160
2" or Less	0.20	0.30	--	--	0.20	--	0.30	--	--	--	0.42
2-1/2"	0.27	0.40	--	--	0.27	--	0.40	--	--	--	0.47
3	0.30	0.42	--	--	0.30	--	0.42	--	--	--	0.58
4	0.42	0.58	--	--	0.42	--	0.58	--	0.84	--	0.90
5	0.54	0.78	--	--	0.54	--	0.78	--	0.91	--	1.07
6	0.70	0.93	--	--	0.70	--	0.93	--	1.18	--	1.35
8	0.97	1.35	0.97	0.97	0.97	1.26	1.35	1.64	1.83	2.04	2.38
10	1.38	1.84	1.38	1.38	1.38	1.84	2.11	2.44	2.58	3.08	3.65
12	1.53	2.44	1.53	1.53	2.07	2.75	2.90	3.17	3.65	4.14	4.38
14 O.D.	2.07	2.67	2.07	2.07	2.67	2.95	3.39	3.85	4.14	--	--
16 O.D.	2.37	3.45	2.37	2.37	3.45	3.71	4.14	4.38	4.87	--	--
18 O.D.	2.95	4.14	2.95	3.65	4.14	4.38	4.87	5.42	--	--	--
20 O.D.	3.54	4.72	3.54	4.72	5.68	6.16	6.49	--	--	--	--
24 O.D.	5.18	6.89	5.18	7.23	7.67	7.93	8.37	--	--	--	--

Above man hours are for die cut IPS pipe threads only.

For threading the ends of bends, add 100% to the above man hours.

WELDED CARBON STEEL ATTACHMENTS

NET MAN HOURS PER LINEAL FOOT

Thickness of Plate Etc. , Inches	Layout & Flame Cutting Per Lineal Inch	Fillet Welding Per Lineal Inch
1/2 or less	. . . . 0. 05	. . . . . . 0. 05
3/4	. . . . 0. 05	. . . . . . 0. 07
1	. . . . 0. 05	. . . . . . 0. 09
1-1/4	. . . . 0. 07	. . . . . . 0. 10
1-1/2	. . . . 0. 07	. . . . . . 0. 10
1-3/4	. . . . 0. 08	. . . . . . 0. 20
2	. . . . 0. 09	. . . . . . 0. 20

Figure labor on basis of total lineal inches to be cut and fillet welded.

Unlisted thickness take the next higher listing.

Any machining of bases, anchors, supports, lugs, etc. , should be charged as an extra.

If preheating is required, add 100% to the above man hours.

DRILLING HOLES IN WELDED ATTACHMENTS

Carbon Steel Material

NET MAN HOURS EACH

Thickness of Plates, Angles, Etc. in Inches	HOLE SIZE			
	3/4" and smaller	7/8", 1" and 1-1/8"	1-1/4", 1-1/2" and 2"	2-1/4" and 2-1/2"
1/2" or less	0.24	0.29	0.34	0.47
3/4	0.29	0.34	0.43	0.64
1	0.31	0.40	0.49	0.67
1-1/4	0.40	0.49	0.55	0.71
1-1/2	0.49	0.55	0.71	0.91
1-3/4	0.55	0.71	0.86	1.12
2	0.71	0.82	1.00	1.32
2-1/2	0.82	0.91	1.12	1.62
3	0.91	1.12	1.32	1.82
3-1/2	1.01	1.21	1.52	2.14
4	1.21	1.42	1.73	2.44

Unlisted thickness of plate or size of holes take the next higher listing.

If holes are to be tapped—Add 33-1/3%.

Drilling of sentinel safety or tell tale holes should be charged at .05 man hours each net.

The above man hours are for drilling holes in flat carbon steel plate and structural shapes only.

For drilling holes in pipe or other contoured objects, perpendicular to contoured surface, add 100% to above man hours.

For drilling holes in pipe or other contoured objects, oblique to contoured surface, add 175% to above man hours.

MACHINING INSIDE OF PIPE
Built-Up-Ends

Carbon Materials Only

Size Inches	Machining Inside of Pipe Net Man Hours per End		Size Inches	Built Up Ends on Inside Diameter of Pipe and Fittings with Weld Metal to Provide for Specified Outside Diameter of Machined Backing Ring
	Standard Extra Strong & Sch. Nos. to 100 Inclusive	Double Extra Strong & Sch.Nos. 120,140 & 160		Net Man Hours per End
2 or less	0.5	0.7	2 or less	0.6
2-1/2	0.5	0.7	2-1/2	0.6
3	0.5	0.7	3	0.7
3-1/2	0.5	0.8	3-1/2	0.7
4	0.7	0.8	4	0.8
5	0.8	0.9	5	0.9
6	0.8	1.0	6	1.0
8	1.0	1.3	8	1.4
10	1.2	1.5	10	2.0
12	1.3	1.8	12	2.5
14 OD	1.5	2.1	14	3.1
16 OD	1.8	2.5	16	3.7
18 OD	2.1	2.8	18	4.6
20 OD	2.5	3.4	20	5.5
24 OD	3.4	4.5	24	8.4

Machining: Man hours for machining the inside of straight pipe are for any taper bore from 10° through 30° included angle. For machining the ends of bends add 100% to the above man hours. For counterboring (up to a maximum of 2″ in length), add 30% to the above man hours. For machining to a controlled "C" dimension (as required for power piping critical systems), add 225% to the above man hours.

Cutting and Beveling: Man hours do not include cutting and beveling. See respective tables for these charges.

Built-Up Ends: Man hours for built-up ends are for building up the I.D. of straight pipe, bends or fittings, at the ends with weld metal and grinding where it is necessary for proper fit of backing rings.

MACHINING INSIDE OF LARGE O.D. PIPE

Built-Up Ends

Carbon Steel Material

O.D. Pipe Size Inches	NET MAN HOURS—PER END Machining Inside of Straight Pipe Only					I.D. Built-up with Weld Material
	WALL THICKNESS IN INCHES					
	.500 to 1.50	1.51 to 2.25	2.26 to 3.00	3.01 to 4.50	4.51 to 6.00	Man Hours Per End
26	4.41	5.30	6.31	8.07	10.11	15.20
28	4.76	5.70	6.79	8.56	10.66	17.98
30	5.30	6.11	7.26	9.10	11.27	22.73
32	5.84	6.79	7.67	9.70	12.01	27.82
34	6.58	7.40	8.41	10.38	12.63	35.01
36	7.40	8.28	9.16	11.00	13.30	41.67
38	8.28	9.23	10.18	11.80	14.04	48.85
40	9.23	10.18	11.27	12.63	14.86	57.06
42	10.24	11.06	12.35	13.44	15.74	66.91
44	11.20	12.28	13.44	14.46	16.56	77.49
46	12.28	13.30	14.58	15.61	17.57	87.32
48	13.44	14.46	15.67	16.70	18.60	98.45
54	15.12	16.26	17.63	18.79	20.92	110.75
60	16.80	18.07	19.59	20.87	23.25	123.06

Machining: Man hours for machining the inside of straight pipe are for any taper bore from 10° through 30° included angle. For machining the ends of bends add 100% to the above man hours. For counterboring (up to a maximum of 2″ in length), add 30% to the above man hours. For machining to a controlled "C" dimension (as required for power piping critical systems), add 225% to the above man hours.

Cutting and Beveling: Man hours do not include cutting and beveling. See respective tables for these charges.

Built-Up Ends: Man hours for built-up ends are for building up the I.D. of straight pipe, bends or fittings, at the ends with weld metal and grinding where it is necessary for proper fit of backing rings.

BORING INSIDE DIAMETER OF PIPE
AND INSTALLING STRAIGHTENING VANES

NET MAN HOURS EACH

Nominal Pipe Size Inches	Boring I.D. of Pipe	Installing Straightening Vanes	
	Carbon Steel	Carbon Steel	Alloy
4	9.8	7.6	11.3
5	11.7	8.7	13.1
6	13.3	10.7	15.2
8	17.5	12.6	18.9
10	20.9	13.9	20.9
12	25.6	15.6	23.7
14	29.5	17.6	26.3
16	35.5	19.5	29.5
18	44.0	22.1	33.0
20	57.7	24.8	37.2
24	79.1	30.2	45.4
26	--	35.9	54.0
28	--	39.6	59.8
30	--	45.9	68.8
32	--	53.1	79.9
34	--	59.8	89.9
36	--	68.8	103.3
38	--	76.9	115.8
40	--	85.2	127.7
42	--	93.7	140.7

Man hours for boring I.D. of pipe include boring for a length of four times nominal pipe size.

Man hours for installing straightening vanes are based on installing vanes in pipe where boring the I.D. of pipe is not required. If boring I.D. of pipe is required or specified, add boring man hours as shown above.

INSTALLING FLOW NOZZLES

Holding Ring Type

Carbon Steel and Alloy Materials

NET MAN HOURS EACH

Pipe Size Inches	FLOW NOZZLES		Pipe O.D. Inches	FLOW NOZZLES	
	Carbon Steel	Alloy		Carbon Steel	Alloy
4	37.8	44.2	26	165.6	198.7
5	41.1	47.3	28	189.6	222.0
6	45.8	53.0	30	217.4	249.9
8	55.2	62.5	32	248.5	282.3
10	63.0	72.7	34	283.7	316.2
12	71.0	82.6	36	319.1	357.4
14 O.D.	77.6	91.2	38	357.5	404.0
16 O.D.	87.6	103.4	40	400.5	456.5
18 O.D.	98.8	117.2	42	448.5	516.0
20 O.D.	111.0	133.7	--	--	--
24 O.D.	140.3	170.7	--	--	--

Man hours include internal machining and nozzle installation.

For installing welding type flow nozzles, add for the bevels, butt weld, butt weld preheat, and any other labor operation or non-destructive testing operation required for the butt weld. See respective tables for these charges.

PREHEATING

Butt Welds and Any Type of Flange Welds

Carbon Steel, or Alloy Materials

For Temperatures Up to 400°F

NET MAN HOURS EACH

Size Ins.	Standard Pipe & OD Sizes 3/8" Thick	Extra Heavy Pipe & OD Sizes 1/2" Thick	SCHEDULE NUMBERS								
			20	30	40	60	80	100	120	140	160
2	0.2	0.3	--	--	0.2	--	0.3	--	--	--	0.5
2-1/2	0.3	0.5	--	--	0.3	--	0.5	--	--	--	0.6
3	0.5	0.6	--	--	0.5	--	0.6	--	--	--	0.7
3-1/2	0.5	0.6	--	--	0.5	--	0.6	--	--	--	0.9
4	0.6	0.7	--	--	0.6	--	0.7	--	0.9	--	0.9
5	0.7	0.9	--	--	0.7	--	0.9	--	0.9	--	1.0
6	0.8	1.0	--	--	0.8	--	1.0	--	1.3	--	1.5
8	0.9	1.3	0.9	0.9	0.9	1.3	1.3	1.8	1.9	2.4	2.5
10	1.3	1.8	1.3	1.3	1.3	1.8	2.0	2.4	2.7	3.3	3.8
12	1.5	2.0	1.5	1.5	1.9	2.2	2.8	3.3	3.8	4.4	5.3
14 OD	1.9	2.5	1.9	1.9	2.2	3.0	3.5	4.4	5.0	5.8	6.6
16 OD	2.2	3.3	2.2	2.2	3.0	3.8	4.5	5.4	6.0	7.3	8.5
18 OD	2.6	3.5	2.6	3.1	4.1	5.0	6.0	7.0	7.9	8.5	10.5
20 OD	3.1	4.1	3.1	4.1	5.2	6.3	7.4	8.7	9.8	11.1	12.9
24 OD	3.7	5.0	3.7	5.3	6.4	7.8	9.3	10.4	11.7	13.3	15.2

Pipe Thickness: The wall thickness of the material determines the man hours that will apply. For preheating of double extra strong material, use schedule 160 man hours.

Mitre Welds: For preheating of mitre welds, add 50% to above man hours.

Man Hours: Man hours for preheating are additional to charges for welding operations.

Preheating: For preheating to temperatures above 400°F. but not exceeding 600°F., add 100% to the above man hours.

PREHEATING HEAVY WALL PIPE BUTT WELDS

Carbon Steel or Alloy Materials

For Temperatures Up to 400°F

NET MAN HOURS EACH

Nominal Pipe Size	WALL THICKNESS IN INCHES							
	.750	1.00	1.25	1.50	1.75	2.00	2.25	2.50
3	1.0	1.2	--	--	--	--	--	--
4	1.4	1.5	1.8	2.0	--	--	--	--
5	--	1.9	2.1	2.4	2.5	2.8	--	--
6	--	2.1	2.5	2.7	3.0	3.2	3.4	--
8	--	3.0	3.4	3.7	3.9	4.4	4.5	4.8
10	--	--	4.1	4.4	4.7	5.4	5.8	6.3
12	--	--	--	6.1	6.6	7.0	7.4	8.0
14	--	--	--	7.3	7.9	8.4	9.1	9.6
16	--	--	--	--	9.4	10.0	10.5	11.6
18	--	--	--	--	--	12.3	13.0	13.7
20	--	--	--	--	--	14.0	15.1	15.9
22	--	--	--	--	--	--	--	17.2
24	--	--	--	--	--	--	--	18.6

	2.75	3.00	3.25	3.50	3.75	4.00	4.25	4.50
10	7.1	7.6	--	--	--	--	--	--
12	8.6	9.1	9.8	10.4	--	--	--	--
14	10.3	10.9	11.6	12.3	13.0	14.0	--	--
16	12.3	13.0	13.7	14.6	15.6	16.6	--	--
18	14.8	15.8	16.8	17.7	18.8	19.8	--	--
20	17.0	18.2	19.4	20.5	21.8	22.9	24.0	25.1
22	18.4	19.8	20.9	22.3	23.8	25.3	26.6	28.0
24	19.8	21.2	22.8	24.2	25.8	27.7	28.4	30.2

	4.75	5.00	5.25	5.50	5.75	6.00
20	26.4	27.7	29.3	30.7	32.1	33.7
22	29.3	30.7	32.1	33.7	35.0	36.5
24	32.1	33.7	34.5	36.5	38.0	39.5

For General Notes, see the bottom of page 121.

PREHEATING LARGE O.D. PIPE BUTT WELDS AND ANY TYPE FLANGE WELDS

Carbon Steel Or Alloy Materials

For Temperatures Up To 400°F

NET MAN HOURS EACH

O.D. Pipe Inches	WALL THICKNESS IN INCHES							
	.500 Or Less	.750	1.00	1.25	1.50	1.75	2.00	2.25
26	7.6	8.5	10.3	12.5	14.5	16.0	18.3	20.5
28	8.3	9.2	10.9	13.1	15.3	17.0	19.5	21.9
30	8.9	9.6	11.8	14.0	16.3	18.1	20.8	23.5
32	9.3	10.3	12.3	14.8	17.2	19.1	21.9	24.8
34	10.0	10.9	13.0	15.8	18.6	20.5	23.2	26.6
36	10.7	11.8	13.8	17.2	20.5	22.5	25.7	29.0
38	10.9	12.5	15.0	18.3	22.8	24.8	28.3	31.9
40	11.2	13.5	16.3	19.5	25.3	27.1	31.0	35.2
42	12.0	14.4	17.6	21.0	28.1	30.0	34.1	38.6
44	13.0	15.3	19.5	24.0	29.3	32.9	37.6	43.3
46	13.9	16.4	21.2	26.3	31.5	36.1	41.1	46.1
48	15.0	17.6	23.0	28.6	33.9	39.3	44.8	50.3
54	16.9	19.8	25.8	32.1	38.1	44.3	50.4	56.5
60	18.8	21.9	28.8	35.6	42.4	49.1	56.1	62.9
	2.50	2.75	3.00	3.25	3.50	3.75	4.00	4.25
26	22.8	24.9	27.1	29.4	31.6	33.7	36.0	38.2
28	24.2	26.3	29.0	30.8	33.0	35.3	37.5	39.8
30	25.7	28.0	30.3	32.1	34.5	36.6	39.3	41.1
32	27.0	29.3	31.9	33.5	35.8	38.0	40.6	42.5
34	28.8	31.0	33.5	35.3	37.6	39.8	42.4	44.3
36	30.2	33.5	36.0	37.9	40.1	42.4	44.8	46.8
	4.50	4.75	5.00	5.25	5.50	5.75	6.00	
26	40.5	42.7	44.8	47.1	49.3	51.6	53.8	
28	42.0	44.0	46.3	48.5	50.7	53.0	55.7	
30	43.3	45.5	47.8	50.0	52.0	54.3	57.0	
32	44.6	46.8	49.1	51.3	53.6	55.8	58.4	
34	46.4	48.0	50.9	53.1	55.0	57.5	60.2	
36	49.0	51.2	53.5	55.7	57.7	59.9	62.7	

For General Notes, see the bottom of page 121.

PREHEATING 90° NOZZLE WELDS

Carbon Steel, or Alloy Materials

For Temperatures Up To 400°F

NET MAN HOURS EACH

Size Ins.	Standard Pipe & OD Sizes 3/8" Thick	Extra Heavy Pipe & OD Sizes 1/2" Thick	SCHEDULE NUMBERS								
			20	30	40	60	80	100	120	140	160
2	0.5	0.6	--	--	0.5	--	0.6	--	--	--	0.7
2-1/2	0.6	0.7	--	--	0.6	--	0.7	--	--	--	0.9
3	0.6	0.9	--	--	0.6	--	0.9	--	--	--	1.0
3-1/2	0.7	0.9	--	--	0.7	--	0.9	--	--	--	--
4	0.9	1.0	--	--	0.9	--	1.0	--	1.4	--	1.7
5	1.0	1.4	--	--	1.0	--	1.4	--	1.7	--	1.8
6	1.3	1.8	--	--	1.3	--	1.8	--	2.0	--	2.4
8	1.8	2.1	1.7	1.7	1.7	2.0	2.1	2.7	3.1	3.5	4.1
10	2.0	2.7	2.0	2.0	2.0	2.7	3.1	3.8	4.5	--	5.9
12	2.5	3.3	2.5	2.5	2.8	3.5	4.5	5.2	6.0	--	8.3
14 OD	3.0	3.8	3.0	3.0	3.5	4.6	5.9	6.6	7.8	--	10.5
16 OD	3.4	4.6	3.4	3.4	4.6	5.9	7.2	8.5	9.8	--	13.7
18 OD	4.2	5.5	4.2	4.8	6.3	7.9	9.6	10.5	12.7	--	17.0
20 OD	5.0	6.5	5.0	6.5	8.3	10.1	11.8	13.7	15.6	--	--
24 OD	6.0	7.8	6.0	8.5	10.3	12.5	15.1	16.5	18.9	--	--

Pipe Thickness: The size of the nozzle and the wall thickness of the header or nozzle (whichever is greater) determines the man hours to be used. For preheating of double extra strong thickness use schedule 160 man hours.

Time: For reinforced 90° nozzle welds, add 100% to the above man hours.
For 45° nozzle welds, add 50% to the above man hours.
For reinforced 45° nozzle welds, add 150% to the above man hours.
For preheating to temperatures above 400°F. but not exceeding 600°F., add 100% to the above man hours.
Preheating of coupling, weldolet, threadolet or socket welds should be charged at the same man hours as shown for the same size and schedule nozzle.
Man hours for preheating are additional to man hours for welding operations.

PREHEATING LARGE O.D. 90° NOZZLE WELDS

Carbon Steel or Alloy Materials

For Temperatures Up To 400°F

NET MAN HOURS EACH

O.D. Pipe Sizes	WALL THICKNESS IN INCHES							
	.500	.750	1.00	1.25	1.50	1.75	2.00	2.25
26	9.7	10.7	12.9	15.6	18.3	20.1	23.0	25.8
28	10.3	11.4	13.8	16.6	19.2	21.2	24.4	27.7
30	10.9	12.2	14.8	17.6	20.5	22.8	26.1	29.5
32	11.6	12.9	15.5	18.5	21.6	24.0	27.5	31.0
34	12.6	13.8	16.3	19.9	23.2	25.7	29.3	33.4
36	13.5	14.8	17.5	21.6	25.7	28.2	32.2	36.5
38	14.3	15.8	18.9	23.0	28.6	31.0	35.5	40.1
40	15.3	17.0	20.2	24.5	31.7	34.2	39.1	44.1
42	16.2	18.1	21.8	26.3	34.0	37.8	43.0	48.5
48	18.5	20.7	24.9	30.1	38.8	43.2	49.1	55.5
54	20.8	23.2	28.1	33.9	43.7	48.5	55.2	62.3
60	23.1	25.8	31.2	37.6	48.5	53.9	61.4	69.3

Pipe Thickness: The size of the nozzle and the wall thickness of the header or nozzle (whichever is greater) determines the man hours to be used. For preheating of double extra strong thickness use schedule 160 man hours.

Time: For reinforced 90° nozzle welds, add 100% to the above man hours.
For 45° nozzle welds, add 50% to the above man hours.
For reinforced 45° nozzle welds, add 150% to the above man hours.
For preheating to temperatures above 400°F. but not exceeding 600°F., add 100% to the above man hours.
Preheating of coupling, weldolet, threadolet or socket welds should be charged at the same man hours as shown for the same size and schedule nozzle.
Man hours for preheating are additional to man hours for welding operations.

LOCAL STRESS RELIEVING

Butt Welds, Nozzle Welds or Any Type of Flange Welds

Carbon Steel Material

Temperatures To 1400°F

NET MAN HOURS EACH

Size Ins.	Standard Pipe & OD Sizes 3/8" Thick	Extra Heavy Pipe & OD Sizes 1/2" Thick	SCHEDULE NUMBERS								
			20	30	40	60	80	100	120	140	160
2	2.6	2.8	--	--	2.6	--	2.8	--	--	--	3.0
2-1/2	2.8	2.9	--	--	2.8	--	2.9	--	--	--	3.1
3	2.9	3.0	--	--	2.9	--	3.0	--	--	--	3.5
3-1/2	3.0	3.1	--	--	3.0	--	3.1	--	--	--	3.8
4	3.0	3.5	--	--	3.0	--	3.5	--	3.6	--	3.9
5	3.5	3.7	--	--	3.5	--	3.7	--	4.1	--	4.3
6	3.7	4.1	--	--	3.7	--	4.1	--	4.3	--	4.9
8	4.2	4.7	4.2	4.2	4.2	4.4	4.7	5.1	5.3	5.5	5.9
10	4.6	5.1	4.6	4.6	4.6	5.1	5.3	5.7	5.9	6.3	6.7
12	5.1	5.5	5.1	5.1	5.3	5.8	6.0	6.5	6.8	7.1	7.4
14 OD	5.5	5.9	5.5	5.5	5.9	6.3	6.7	7.1	7.6	7.9	8.3
16 OD	5.9	6.4	5.9	5.9	6.4	6.8	7.2	7.8	8.0	8.5	9.2
18 OD	6.4	6.8	6.4	6.6	6.8	7.3	7.8	8.3	8.7	9.2	10.1
20 OD	6.6	7.0	6.6	6.8	7.3	7.8	8.3	9.2	9.6	10.0	11.1
24 OD	7.1	7.3	7.1	7.6	8.0	8.5	9.2	10.1	10.5	11.2	12.5

Pipe Thickness: For stress relieving butt welds and flange welds, the wall thickness of the pipe determines the man hours that will apply. For stress relieving nozzle welds, the size and thickness of the header to which the nozzle is attached determines the man hours that will apply. For local stress relieving of double extra strong material, use schedule 160 man hours.

Code Requirements: All welds in piping materials having a wall thickness of 3/4" or greater must be stress relieved to comply with the requirements of the A.S.A. Code for pressure piping. Man hours shown below the ruled line in the above schedule cover sizes having a wall thickness of 3/4" or greater.

HEAVY WALL LOCAL STRESS RELIEVING

Butt Welds

Carbon Steel Material

Temperatures To 1400°F

NET MAN HOURS EACH

Nominal Pipe Size	WALL THICKNESS IN INCHES							
	.750	1.00	1.25	1.50	1.75	2.00	2.25	2.50
3	5.1	5.4	--	--	--	--	--	--
4	5.4	5.9	6.4	6.8	--	--	--	--
5	--	6.3	6.7	7.2	7.6	8.2	--	--
6	--	6.7	7.2	7.8	8.2	9.0	9.7	--
8	--	7.4	7.8	8.4	8.9	9.5	10.1	10.8
10	--	--	8.1	8.8	9.2	9.8	10.3	11.2
12	--	--	--	9.0	9.6	10.3	10.8	11.5
14	--	--	--	9.6	10.3	11.2	11.6	12.3
16	--	--	--	--	10.8	11.6	12.3	13.1
18	--	--	--	--	--	12.3	13.1	14.0
20	--	--	--	--	--	13.3	14.3	15.2
22	--	--	--	--	--	--	--	16.4
24	--	--	--	--	--	--	--	17.7
	2.75	3.00	3.25	3.50	3.75	4.00	4.25	4.50
10	11.9	12.5	--	--	--	--	--	--
12	12.3	13.2	14.0	15.0	--	--	--	--
14	13.2	14.1	15.0	16.1	16.9	18.1	--	--
16	14.0	15.0	16.0	16.9	18.1	19.3	--	--
18	15.0	16.0	16.9	18.3	19.4	20.6	--	--
20	16.2	17.4	18.4	19.6	21.2	22.7	24.3	25.9
22	17.4	19.2	20.0	21.3	22.8	24.3	25.9	27.4
24	18.9	20.1	21.5	22.9	24.5	26.0	27.6	29.1
	4.75	5.00	5.25	5.50	5.75	6.00		
20	27.4	29.0	30.5	32.1	33.7	35.3		
22	29.0	30.6	32.1	33.7	35.3	36.9		
24	30.6	32.2	33.7	35.3	36.8	38.4		

For General Notes, see the bottom of page 126.

LARGE O.D. LOCAL STRESS RELIEVING

Butt Welds, Nozzle Welds or Any Type Flange Weld

Carbon Steel Material

Temperatures To 1400°F

NET MAN HOURS EACH

O.D. Pipe Size	WALL THICKNESS IN INCHES								
	.375	.500	.750	1.00	1.25	1.50	1.75	2.00	2.25
26	9.1	9.8	11.3	11.8	12.7	14.1	16.3	18.5	21.3
28	9.5	10.5	12.2	12.7	13.8	15.2	17.7	20.0	22.9
30	10.4	11.3	13.1	13.9	14.7	16.7	19.2	21.5	24.6
32	11.2	12.3	14.3	15.0	15.8	18.1	20.8	23.2	26.5
34	12.4	13.5	15.5	16.2	17.1	19.6	22.5	24.8	28.2
36	13.6	14.7	17.0	18.1	19.3	21.5	24.5	27.0	30.6
38	14.7	16.1	18.7	20.0	21.6	23.9	27.0	29.3	33.5
40	16.1	17.6	20.6	22.3	24.3	26.6	29.7	32.0	36.5
42	17.8	19.4	22.7	24.6	27.3	29.6	32.5	35.0	39.7
44	19.9	21.5	24.7	27.0	29.7	32.5	35.8	39.2	42.8
46	21.9	23.6	26.9	29.9	32.7	36.0	39.3	42.8	46.5
48	24.4	26.0	29.2	32.7	35.8	39.3	42.9	46.3	49.8
54	28.5	29.2	32.9	36.8	40.3	45.4	49.6	53.5	57.5
60	30.4	32.4	36.6	41.9	44.7	49.2	53.6	58.0	63.6

	2.50	2.75	3.00	3.25	3.50	3.75	4.00	4.25	4.50
26	23.8	26.5	29.1	31.7	34.4	37.0	39.7	42.3	45.0
28	25.5	28.2	30.8	33.5	36.1	38.8	41.4	44.0	46.6
30	27.3	29.9	32.5	35.2	37.8	40.4	43.0	45.7	48.3
32	29.1	31.7	34.3	36.9	39.6	42.2	44.9	47.5	50.1
34	30.8	33.5	36.1	38.0	41.4	44.0	46.7	49.3	52.0
36	33.2	35.9	37.3	40.4	43.8	46.5	49.1	51.8	54.4

	4.75	5.00	5.25	5.50	5.75	6.00			
26	47.6	50.3	52.8	55.4	58.1	60.7			
28	49.2	51.9	54.5	57.2	59.8	62.4			
30	50.9	53.6	56.2	58.8	61.4	64.1			
32	52.8	55.4	58.1	60.6	63.3	65.9			
34	54.6	57.3	59.9	62.4	65.1	67.7			
36	56.9	59.6	62.2	64.9	67.5	70.2			

For General Notes, see the bottom of page 126.

RADIOGRAPHIC INSPECTION

Field X-Ray Of Butt Welds

Carbon Steel Material

NET MAN HOURS EACH

Nominal Pipe Size	Wall Thickness Thru Extra Strong	Wall Thickness Greater Than Extra Strong Thru Schedule 120	Wall Thickness Greater Than Schedule 120 Thru Double Extra Strong
2 Or Less	0.86	--	1.13
3	0.86	--	1.13
4	0.98	1.13	1.27
5	1.07	1.23	1.40
6	1.20	1.40	1.56
8	1.35	1.54	1.75
10	1.51	1.73	1.97
12	1.71	1.97	2.23
14	1.86	2.14	2.42
16	2.08	2.39	2.70
18	2.32	2.67	3.01
20	2.55	2.94	3.34
24	3.15	3.62	4.08

Man hours listed above cover radiographic inspection of butt welded joints by x-raying.

For radiographic inspection of mitre butt welds, add 50% to above man hours.

For radiographic inspection of slip-on flange welds, add 100% to above man hours.

For radiographic inspection of nozzle welds add 200% to above man hours.

HEAVY WALL RADIOGRAPHIC INSPECTION

Field X-Ray Of Butt Welds

Carbon Steel Material

NET MAN HOURS EACH

Nominal Pipe Size	WALL THICKNESS IN INCHES							
	.750	1.00	1.25	1.50	1.75	2.00	2.25	2.50
3	1.24	1.37	--	--	--	--	--	--
4	1.37	1.43	1.46	1.65	--	--	--	--
5	--	1.50	1.59	1.72	1.81	1.97	--	--
6	--	1.59	1.72	1.81	1.97	2.09	2.25	--
8	--	1.74	1.87	2.00	2.16	2.28	2.46	2.75
10	--	--	2.08	2.19	2.40	2.50	2.69	2.96
12	--	--	--	2.38	2.59	2.74	2.94	3.12
14	--	--	--	2.62	2.78	3.02	3.21	3.40
16	--	--	--	--	3.03	3.24	3.44	3.69
18	--	--	--	--	--	3.53	3.80	4.02
20	--	--	--	--	--	3.87	4.10	4.40
22	--	--	--	--	--	--	--	4.90
24	--	--	--	--	--	--	--	5.41
	2.75	3.00	3.25	3.50	3.75	4.00	4.25	4.50
10	3.16	3.33	--	--	--	--	--	--
12	3.33	3.59	3.80	4.06	--	--	--	--
14	3.63	3.87	4.10	4.40	4.68	5.00	--	--
16	3.97	4.21	4.46	4.81	5.11	5.41	--	--
18	4.33	4.62	4.93	5.21	5.58	5.98	--	--
20	4.68	5.00	5.31	5.65	6.04	6.44	6.87	7.28
22	5.22	5.58	5.97	6.33	6.74	7.24	7.73	8.20
24	5.81	6.20	6.62	7.00	7.51	8.01	8.76	9.11
	4.75	5.00	5.25	5.50	5.75	6.00		
20	7.73	8.17	8.58	9.01	9.40	9.83		
22	8.54	8.95	9.36	9.77	10.03	10.61		
24	9.45	9.77	10.13	10.60	11.04	11.49		

For General Notes, see the bottom of page 129.

LARGE O.D. RADIOGRAPHIC INSPECTION

Field X-Ray Of Butt Welds
Carbon Steel Material

NET MAN HOURS EACH

O.D. Pipe Size	WALL THICKNESS IN INCHES							
	.750 Or Less	1.00	1.25	1.50	1.75	2.00	2.25	2.50
26	3.62	3.81	3.96	4.27	4.57	4.91	5.24	5.64
28	4.20	4.35	4.53	4.82	5.13	5.46	5.83	6.23
30	5.06	5.24	5.39	5.68	6.01	6.33	6.66	7.11
32	6.26	6.41	6.56	6.88	7.21	7.52	7.82	8.25
34	7.75	7.94	8.07	8.40	8.71	9.02	9.35	9.78
36	9.53	9.78	9.88	10.19	10.51	10.83	11.19	11.56
38	11.37	11.57	11.78	12.08	12.40	12.72	13.09	--
40	13.36	13.64	13.83	14.11	14.50	14.85	15.20	--
42	15.53	15.76	15.97	16.32	16.72	17.04	17.41	--
44	17.87	18.06	18.31	18.65	19.08	19.46	19.80	--
46	20.27	20.52	20.79	21.16	21.59	21.93	22.26	--
48	22.84	23.07	23.33	23.70	24.20	24.50	24.86	--
54	25.69	25.96	26.25	26.66	27.22	27.55	27.97	--
60	28.54	28.83	29.16	29.62	30.25	30.61	31.07	--

	2.75	3.00	3.25	3.50	3.75	4.00	4.25	4.50
26	6.11	6.68	7.15	7.54	8.12	8.61	9.57	10.00
28	6.72	7.52	7.80	8.15	8.72	9.32	10.19	10.60
30	7.54	8.43	8.65	9.02	9.59	10.19	11.02	11.44
32	8.67	9.57	9.79	10.15	10.73	11.34	12.18	12.58
34	10.24	11.07	11.34	11.68	12.26	12.85	13.71	14.13
36	12.05	12.90	13.13	13.49	14.05	14.66	15.53	15.95

	4.75	5.00	5.25	5.50	5.75	6.00
26	10.35	10.65	10.95	11.41	11.71	12.14
28	10.95	11.28	11.56	11.99	12.29	12.75
30	11.83	12.12	12.42	12.85	13.18	13.62
32	12.97	13.27	13.56	13.98	15.12	15.61
34	14.48	14.81	15.07	15.53	16.68	17.55
36	16.31	16.62	16.89	17.33	18.47	18.23

For General Notes, see the bottom of page 129.

HYDROSTATIC TESTING

For Pressures Not Exceeding 4,000 P.S.I.

Carbon Steel Material

NET MAN HOURS PER LINEAR FOOT

Pipe Size Inches	Standard Pipe & O.D. Sizes 3/8" Thick	Extra Hvy. Pipe & O.D. Sizes 1/2" Thick	SCHEDULE NUMBERS								
			20	30	40	60	80	100	120	140	160
2" Or Less	0.014	0.016	--	--	0.014	--	0.016	--	--	--	0.025
2-1/2"	0.015	0.017	--	--	0.015	--	0.017	--	--	--	0.028
3	0.017	0.019	--	--	0.017	--	0.019	--	--	--	0.031
4	0.020	0.024	--	--	0.020	--	0.024	--	0.032	--	0.036
5	0.022	0.026	--	--	0.022	--	0.026	--	0.035	--	0.041
6	0.025	0.029	--	--	0.025	--	0.029	--	0.039	--	0.047
8	0.027	0.032	0.027	0.027	0.027	0.030	0.032	0.039	0.046	0.052	0.057
10	0.031	0.035	0.031	0.031	0.031	0.035	0.041	0.049	0.055	0.063	0.070
12	0.034	0.039	0.034	0.034	0.038	0.044	0.053	0.061	0.068	0.077	0.087
14 O.D.	0.038	0.044	0.038	0.038	0.041	0.050	0.062	0.069	0.078	0.091	0.106
16 O.D.	0.044	0.049	0.044	0.044	0.049	0.062	0.075	0.086	0.097	0.114	0.134
18 O.D.	0.051	0.058	0.051	0.055	0.064	0.077	0.095	0.108	0.123	0.141	0.164
20 O.D.	0.057	0.067	0.057	0.067	0.078	0.095	0.115	0.133	0.151	0.172	0.203
24 O.D.	0.076	0.086	0.076	0.087	0.115	0.141	0.175	0.200	0.230	0.265	0.304

Above man hours are average for testing completed process systems, for a maximum holding time of one hour, and include time for the following operations when required:

1) Place and remove blinds and blanks as required.
2) Opening and closing of valves.
3) Removal and replacement of valves, orifice plates, expansion joints, and short pieces of pipe as may be required.
4) Block up and block removal of spring-supported or counterweight-supported lines.
5) Air purging of lines before hydro-test.
6) Soap testing joints where required.
7) Drain lines after testing.

If individual segments or spools are to be tested separately multiply above man hours by a factor of ten (10).

HYDROSTATIC TESTING—HEAVY WALL PIPE

For Pressures Not Exceeding 4,000 P.S.I.

Carbon Steel Material

NET MAN HOURS PER LINEAR FOOT

Nominal Pipe Size	WALL THICKNESS IN INCHES							
	.750	1.00	1.25	1.50	1.75	2.00	2.25	2.50
3	0.035	0.039	--	--	--	--	--	--
4	0.040	0.042	0.044	0.048	--	--	--	--
5	--	0.046	0.049	0.052	0.059	0.072	--	--
6	--	0.053	0.058	0.060	0.078	0.091	0.106	--
8	--	0.064	0.068	0.076	0.097	0.113	0.133	0.160
10	--	--	0.076	0.095	0.118	0.136	0.160	0.188
12	--	--	--	0.103	0.122	0.143	0.170	0.196
14	--	--	--	0.117	0.139	0.163	0.191	0.225
16	--	--	--	--	0.156	0.186	0.217	0.260
18	--	--	--	--	--	0.208	0.242	0.286
20	--	--	--	--	--	0.242	0.286	0.346
22	--	--	--	--	--	--	--	0.378
24	--	--	--	--	--	--	--	0.411
	2.75	3.00	3.25	3.50	3.75	4.00	4.25	4.50
10	0.192	0.220	--	--	--	--	--	--
12	0.224	0.257	0.289	0.330	--	--	--	--
14	0.256	0.290	0.329	0.372	0.424	0.476	--	--
16	0.294	0.338	0.381	0.433	0.494	0.563	--	--
18	0.328	0.381	0.433	0.497	0.567	0.649	--	--
20	0.394	0.442	0.506	0.571	0.649	0.736	0.834	0.905
22	0.430	0.485	0.552	0.623	0.709	0.805	0.913	1.010
24	0.468	0.517	0.598	0.676	0.770	0.875	0.992	1.094
	4.75	5.00	5.25	5.50	5.75	6.00		
20	0.992	1.062	1.133	1.180	1.314	1.400		
22	1.070	1.149	1.257	1.320	1.446	1.542		
24	1.168	1.257	1.374	1.440	1.558	1.666		

For General Notes, see the bottom of page 132.

HYDROSTATIC TESTING—LARGE O.D. PIPE

For Pressures Not Exceeding 4,000 P.S.I.

Carbon Steel Material

NET MAN HOURS PER LINEAR FOOT

O.D. Pipe Inches	WALL THICKNESS IN INCHES							
	.500 or Less	.750	1.00	1.25	1.50	1.75	2.00	2.25
26	0.140	0.161	0.185	0.213	0.245	0.281	0.323	0.389
28	0.162	0.186	0.214	0.246	0.283	0.326	0.375	0.431
30	0.194	0.223	0.257	0.296	0.340	0.391	0.450	0.518
32	0.241	0.277	0.319	0.367	0.422	0.485	0.558	0.642
34	0.299	0.344	0.396	0.455	0.523	0.601	0.691	0.795
36	0.368	0.423	0.487	0.560	0.644	0.741	0.852	0.980
38	0.438	0.504	0.580	0.667	0.767	0.882	1.014	1.166
40	0.517	0.595	0.684	0.787	0.905	1.041	1.197	1.377
42	0.600	0.690	0.794	0.913	1.050	1.208	1.389	1.597
44	0.690	0.794	0.913	1.050	1.208	1.389	1.597	1.837
46	0.780	0.897	1.032	1.187	1.365	1.570	1.806	2.077
48	0.881	1.013	1.165	1.340	1.541	1.772	2.038	2.344
54	0.987	1.135	1.305	1.501	1.726	1.985	2.283	2.626
60	1.096	1.260	1.449	1.666	1.916	2.203	2.534	2.914

	2.50	2.75	3.00	3.25	3.50	3.75	4.00	4.25
26	0.447	0.509	0.562	0.650	0.735	0.837	0.951	1.078
28	0.492	0.561	0.617	0.716	0.809	0.922	1.051	1.188
30	0.551	0.628	0.691	0.802	0.906	1.033	1.178	1.331
32	0.738	0.841	0.925	1.073	1.213	1.383	1.577	1.782
34	0.914	1.042	1.146	1.329	1.502	1.712	1.952	2.206
36	1.127	1.285	1.414	1.640	1.853	2.112	2.408	2.721

	4.50	4.75	5.00	5.25	5.50	5.75	6.00	
26	1.189	1.269	1.366	1.493	1.565	1.693	1.810	
28	1.260	1.348	1.456	1.587	1.666	1.799	1.925	
30	1.361	1.456	1.573	1.715	1.801	1.945	2.081	
32	1.960	2.097	2.265	2.469	2.593	2.800	2.996	
34	2.427	2.597	2.805	3.058	3.211	3.468	3.711	
36	2.993	3.203	3.459	3.770	3.959	4.276	4.575	

For General Notes, see the bottom of page 132.

ACCESS HOLES

Carbon Steel Material

NET MAN HOURS EACH

Nominal Pipe Size	WALL THICKNESS				
	Up To 1"	OVER 1" to 2"	OVER 2" to 2-1/2"	OVER 2-1/2" to 4"	OVER 4" to 6"
2-1/2, 3, 4	1.8	2.0	--	--	--
5, 6, 8	2.0	2.2	2.6	--	--
10, 12	2.2	2.4	2.9	4.0	--
14, 16, 18	2.3	2.5	3.0	4.3	--
20, 22, 24	2.5	2.9	3.1	4.5	6.8
26, 28, 30	2.9	3.2	3.3	4.8	7.2
32, 34, 36	3.1	3.3	3.7	5.1	7.6
38, 40, 42	3.3	3.7	4.0	5.6	8.4
44, 46, 48	3.7	4.0	4.4	6.7	10.1
54, 60	4.6	4.9	5.4	8.3	12.5

Man hours include access holes through 1" diameter (drilled and tapped) for radiographic inspection of welds when specified or required.

For openings larger than 1" in diameter, add 25% to the above man hours for each ¼" increase in diameter.

If plugs are to be included and seal welded, add 0.75 man hours each.

INSTRUMENT AND CONTROL PIPING

TRACING

Man hours are for all fabrication and installation or erection of all copper tubing, steel piping, alloy tubing and capillary tubing used for tracing or for interconnecting all instruments and control devices, regardless of location, whether on panel board, piping or equipment, for all instrument and control piping up to and including one (1) inch.

Man hours include the labor of installing all connections, fittings, a n d supporting clips or hangers.

Should the lines be group installed and protected by casings or channels, the installation labor of such casings or channels are extra.

Runs up to 10 feet . . . .   5. 0 Man hours total

Runs over 10 feet . . . .   0. 4 Man hours per foot of length

Asbestos wrap pipe before tracing . . . .   0. 05 Man hours per foot

---

SOLDERED NON-FERROUS FITTINGS

NET MAN HOURS EACH

Nominal Size—Ins.	Couplings	Ells	Tees	Flanges	Reducers	Adapters	Unions	Caps & Plugs	Valves
1/8–3/8	.16	.17	.24	.16	.18	.15	.18	.08	.35
1/2	.22	.23	.33	.22	.25	.20	.25	.11	.40
3/4	.28	.29	.42	.28	.32	.25	.35	.14	.45
1"	.43	.44	.63	.32	.50	.40	.55	.22	.60
1-1/4	.80	.85	1.20	.64	.85	.75	.90	.40	1.00
1-1/2	.85	.90	1.26	.70	.90	.80	1.00	.42	1.15
2	.96	1.00	1.44	.90	1.00	.92	1.15	.50	1.25
2-1/2	1.50	1.55	2.25	1.15	1.65	1.35	1.85	.75	2.10
3	1.95	2.00	2.91	1.40	2.00	1.80	2.25	.95	2.50
4	2.00	2.25	3.00	1.65	2.15	1.90	2.75	1.00	3.00

Man hours include procuring, handling, and complete jointing or making on of solder-type cast and wrought pressure type brass or copper fittings.

Man hours do not include installation of tubing or pipe, supports, instruments or scaffolding. See respective tables for these charges.

PVC-PLASTIC PIPE

MAN HOURS PER UNITS LISTED

Size Ins.	Handle Pipe Per L. F.	Cemented Socket Joints — Ea.	Saddles Each	Handle Valves PCV Body — Ea.
1/2	.07	.20	.38	.13
3/4	.07	.22	.39	.16
1	.07	.25	.40	.17
1-1/4	.08	.27	.43	.20
1-1/2	.08	.29	.45	.25
2	.09	.33	.50	.35
2-1/2	.09	.38	.55	.58
3	.10	.45	.63	.85
4	.11	.55	.73	1.25
6	.12	.70	.90	1.60
8	.14	.80	1.00	1.95
10	.17	1.00	1.20	2.50
12	.20	1.25	1.40	3.00

Handle Pipe Units: Man hours include handling, hauling rigging and aligning in place.

Cement Socket Joint Units: Man hours include cut, square, ream, fit-up and make joint.

Saddle Units: Man hours include fit-up, drill hole in header and cement saddle to header. Maximum hole size is assumed to be 1½ inch. For larger branch lines the use of tees should be estimated. The size of the header not the size of the saddle determines the man hours that apply.

Handle Valve Units: Man hours include handling, hauling and positioning of valve only. Connections of the type as required must be added.

Units are for all wall thickness.

Units do not include scaffolding. See respective table for this charge.

SARAN LINED STEEL PIPE AND FITTINGS

MAN HOURS PER UNITS LISTED

Size Inches	Pipe Per L.F.	Cut & Thread Each	Make-Ons Each	Bolt-Ups Each
1	.20	.10	.20	.50
1-1/4	.22	.15	.30	.60
1-1/2	.23	.16	.35	.65
2	.25	.25	.40	.70
2-1/2	.27	--	--	.80
3	.30	--	--	.85
4	.35	--	--	1.20
6	.40	--	--	1.50
8	.52	--	--	2.10

Pipe Units: Pipe units include rigging, erecting and aligning of pipe. Pipe is normally furnished in 10-foot lengths, with ends threaded and flanges installed at the factory. If this is the case, add 15 percent to above pipe man hours for handling additional weight.

Make-On Units: Make-on units include making on of screwed type fittings. Ells = two make-ons, tees = three make-ons, etc.

Bolt-Up Units: Bolt-up units include bolting together of flanged connections.

All Units: All units include unloading, handling and hauling to storage and erection site.

Man hours do not include supports or scaffolding. See respective tables for these charges.

SCHEDULE 30 OR 40 RUBBER-LINED
STEEL PIPE AND FITTINGS

MAN HOURS PER UNITS LISTED

Size Ins.	Pipe Per L. F.	Cut & Thread Each	Make-Ons Each	Bolt-Ups Each
1-1/4	.22	.15	.30	.60
1-1/2	.23	.16	.35	.65
2	.25	.25	.40	.70
2-1/2	.27	.30	.45	.80
3	.30	.32	.50	.85
3-1/2	.33	.35	.55	1.00
4	.35	.50	.60	1.20
5	.38	--	--	1.30
6	.40	--	--	1.50
8	.52	--	--	2.10
10	.60	--	--	2.70
12	.75	--	--	3.40

Pipe Units: Pipe units include rigging, erecting and aligning of pipe.

Cut Thread Units: Cut and thread units include hand cutting and threading of pipe two inches and smaller and machine cutting and threading of size two and one half inches and larger.

Make-On Units: Make-on units include making on of screwed type fittings. Ells = two make-ons, tees = three make-ons, etc.

Bolt-Up Units: Bolt-up units include bolting together of flanged connections.

All Units: All units include unloading, handling and hauling to storage and erection site.

If pipe is received from factory in flanged 20 foot lengths add 10 percent to pipe handling manhours for handling of additional weight.

Man hours do not include supports or scaffolding. See respective tables for these charges.

SCHEDULE 40 LEAD LINED STEEL PIPE AND FITTINGS

MAN HOURS PER UNITS LISTED

Pipe Size	Handle Pipe Per L. F.	Cut and Thread Each	Make-Ons Each	Butt Welds Incl. Lead Burning Each	Bolt-Ups Each
1-1/4	.22	.15	.30	.90	.60
1-1/2	.25	.16	.35	1.10	.65
2	.33	.25	.38	1.30	.70
2-1/2	.35	.30	.40	1.45	.80
3	.40	.33	.45	1.55	.85
3-1/2	.45	.35	.47	1.65	1.00
4	.50	.50	.55	1.85	1.20
5	.65	--	--	2.20	1.35
6	.80	--	--	2.80	1.50
8	1.20	--	--	3.40	2.10
10	1.60	--	--	4.30	2.70
12	2.25	--	--	5.55	3.40

Pipe Units: Pipe units include rigging, erecting and aligning of pipe. It is customary to order this type piping prefabricated with lead lined fittings in place. Should this be the case add 20 percent to handling pipe units for elimination of field make-ons and handling of additional weight.

Cut and Thread Units: Cut and thread units include hand cutting and threading of pipe two inches and smaller and machine cutting and threading of sizes two and one half inches and larger.

Make-On Units: Make-on units include making on of screwed type fittings. Ells = two make-ons, tees = three make-ons, etc.

Butt Welds: Butt welds including lead burning units include circumfemtial manual electric weld and fusing together of lead at joint.

Bolt-Up Units: Bolt-up units include bolting together of flanged connections.

All Units: All units include unloading, handling and hauling to storage and erection site.

Man hours do not include supports or scaffolding. See respective tables for these charges.

FLANGED CAST IRON CEMENT LINED PIPE AND FITTINGS

MAN HOURS PER UNITS LISTED

Pipe Size Ins.	Handle Pipe Per Foot	Handle Fittings Each	Handle Valves Each	Flange Bolt-Up Each
3	.20	.43	1.20	.80
4	.27	.61	1.70	1.20
6	.40	.79	2.20	1.50
8	.50	1.00	2.80	2.10
10	.65	1.30	3.60	2.70
12	.80	1.55	4.30	3.40
14	.95	1.84	5.10	3.80
16	1.25	2.12	5.90	4.40

Pipe, fittings, and valve units include handling, unloading, hauling to storage and erection sites, and setting and aligning.

Bolt-up man hours include bolting up of flanged joints.

Above man hours are for installation on pipe racks to 20′0″ high. For other installation conditions add or deduct the following percentage.

On sleepers 2′0″ high .... Deduct 12%
In enclosed passage way .... Add 80%
Inside buildings (horizontal or vertical) .... Add 60%
In battery limits of process area .... Add 150%

SCHEDULE 40 CEMENT LINED CARBON STEEL PIPE WITH STANDARD FITTINGS

MAN HOURS PER UNITS LISTED

Pipe Size Ins.	Handle Pipe Per Foot	Cutting Pipe Per Cut	Butt Welds Each	Sleeve Joint With Two Welds Each	90° Welded Nozzle Each	Smooth On Cement Per Joint
6	.40	.65	2.0	3.30	6.05	.35
8	.50	1.00	2.6	4.60	7.30	.50
10	.65	1.20	3.1	5.70	8.30	.60
12	.80	1.45	4.1	6.90	11.35	.65
14	.95	2.30	5.0	7.90	13.90	.80
16	1.25	2.95	6.6	9.20	18.15	.95
18	1.40	3.70	8.6	10.40	22.80	1.10
20	1.75	4.65	9.4	12.40	26.95	1.25
24	2.20	5.90	13.3	15.50	33.60	1.50

Handle Pipe Units: Man hours include handling, unloading, hauling to storage and erection site and setting and aligning.

Cutting Pipe Units: Man hours include cutting pipe and lining. Lining to be cut square and flush with ends.

Butt Weld Units: Man hours include making complete electric weld. Cement lining should be wet with water around welding area.

Sleeve Joint Units: Man hours include slipping on of sleeve, aligning and welding at both ends.

90° Welded Nozzle Units: Man hours include complete operations for welding nozzle.

Smooth on Cement Units: Man hours include mixing and patching weld joints with cement.

Man hours do not include excavation or racks or supports. See respective tables for these charges.

DOUBLE TOUGH PYREX PIPE AND FITTINGS

MAN HOURS PER UNITS LISTED

Pipe Size	Erect Spool Piece Pipe Per L. F.	Standard Bolt-Up Each	Split Flange Bolt-Up Each
1	.35	.35	.50
1-1/2	.45	.40	.70
2	.55	.45	.80
3	.66	.55	.90
4	.75	.85	1.33
6	1.00	.95	1.50
8	1.30	1.00	2.00

Pyrex pipe is usually ordered from factory prefabricated into spool pieces with all necessary fittings and standard flanges in place. Should pipe be ordered without factory installed flanges it is good practice to install split type flanges in the field.

Above man hours are based on installing factory fabricated spool pieces, up to 10 feet long, in the field and include all handling, unloading and hauling to storage and erection site.

Spool Piece Units: Spool piece units include rigging, erecting and aligning.

Standard Bolt-Up Units: Standard bolt-up units include bolting up of factory furnished fabricated flanges.

Split Flange Bolt-Up Units: Split flange bolt-up units include all operations necessary for the slipping on of flanges and gaskets and bolting-up.

Man hours do not include installation of hangers or supports or the erection of scaffolding. See respective tables for these charges.

OVERHEAD TRANSITE
PRESSURE PIPE–CLASS 150

NET MAN HOURS PER UNITS LISTED

Size Inches	Pipe Per L. F.	Make-Ons Each	Bolt-Ups Each
4	.20	.50	1.2
6	.25	.75	1.5
8	.30	.85	2.1
10	.40	1.00	2.7
12	.50	1.25	3.4
14	.60	1.50	3.8
16	.70	1.75	4.4

Pipe Units: Pipe units include rigging, erecting and aligning of pipe.

Make-On Units: Make-on units include erecting, aligning pouring joint. Ells = two make-ons, tees = three make-ons, etc.

Bolt-Up Units: Bolt-up units include bolting together of flanged connections.

All Units: All units include unloading, handling and hauling to storage and erection site.

Man hours do not include supports or scaffolding. See respective tables for these charges.

Transite pipe = 4 inches I.D. and above. It is usually supplied in standard 13-foot lengths.

Section Three

ALLOY
AND
NON-FERROUS
FABRICATION

This section is intended to cover the complete shop fabrication and field fabrication and erection of alloy and non-ferrous piping. Alloy and non-ferrous piping operations are to be man houred on the same basis as corresponding carbon steel materials, *plus* the percentage given in the following tables applicable to the carbon steel man hours listed in sections one and two.

Fabrication of alloy and non-ferrous piping is to be figured on the same basis as for corresponding operation on carbon steel materials, plus the percentage given below applicable to the carbon steel man hour schedules.

SHOP HANDLING PIPE FOR FABRICATION

PERCENT ADDITIVE

| Nominal Pipe Size | MATERIAL CLASSIFICATION-GROUP NUMBERS AND PERCENTAGES | | | | | | | | |
	Group 1	Group 2	Group 3	Group 4	Group 5	Group 6	Group 7	Group 8	Group 9
2 or less	17.0	22.5	30.0	31.8	13.0	39.0	84.0	36.0	46.0
3	18.5	25.0	33.0	35.0	15.0	41.0	88.0	39.0	51.0
4	20.0	27.0	36.0	38.0	17.0	45.5	90.5	41.0	55.0
5	21.0	28.0	38.0	40.0	18.0	49.0	—	42.0	58.0
6	23.0	31.0	41.5	42.5	20.0	50.0	94.0	43.5	64.0
8	26.0	35.0	47.0	48.0	33.5	59.0	100.5	49.5	72.0
10	28.5	38.0	51.0	54.0	50.0	64.0	117.0	57.0	78.0
12	30.0	40.0	54.0	58.0	54.0	70.0	134.0	64.0	83.0
14	33.0	44.0	59.0	62.0	67.0	78.0	--	67.0	90.0
16	35.0	47.0	63.0	67.0	74.0	86.0	--	77.0	96.0
18	39.5	53.0	71.0	75.0	77.0	92.0	--	82.0	108.5
20	43.5	58.0	78.0	83.0	84.0	100.0	--	87.0	120.0
22	46.0	62.0	83.0	88.0	89.0	103.5	--	93.0	127.0
24	49.0	66.0	88.0	93.0	94.0	107.0	--	97.0	135.0
26	53.0	71.5	95.4	100.1	101.9	116.0	—	105.0	146.4
28	57.0	77.0	102.8	108.6	109.8	124.9	—	113.0	157.6
30	61.0	82.5	110.0	116.4	117.6	133.8	—	121.0	168.9
32	65.3	88.0	117.4	124.2	125.4	142.7	—	129.3	180.2
34	69.4	93.5	124.8	131.9	133.3	151.6	—	137.4	191.4
36	73.4	99.0	132.0	139.7	141.1	160.6	—	145.4	202.7
38	77.5	104.5	139.5	147.4	149.0	169.5	—	153.5	213.9
40	81.6	110.0	146.8	155.2	156.8	178.4	—	161.6	225.2
42	85.7	115.5	154.0	163.0	164.6	187.3	—	169.7	236.5

GROUP 1	ASTM A335-P1	.50% Moly
	ASTM A335-P2	.50-.70% Chr.
	ASTM A335-P12	.85-1.10% Chr.
	ASTM A335-P11	1.05-1.45% Chr.
	ASTM A335-P3	1.50-2.00% Chr.

GROUP 2	ASTM A335-P3b	1.75-2.25% Chr.
	ASTM A335-P22	2.00-2.50% Chr.
	ASTM A335-P21	2.75-3.25% Chr.
	ASTM A335-P5,b,c	4.00-6.00% Chr.

GROUP 3	ASTM A335-P7	6.00-8.00% Chr.
	ASTM A335-P9	8.00-10.00% Chr.
	Ferritic Chromes	10.00-15.00% Chr.
	ASTM A333 Gr. 3	3.50% Nickel

| GROUP 4 | Stainless Steels
Types 304, 309, 310, 316
(including "L" & "H" Grades) |
| GROUP 5 | Copper, Brass, Everdur |

| GROUP 6 | ASTM A-333-GR-1
ASTM A-333-GR4
ASTM A-333-GR9 |

GROUP 7	Hastelloy, Titanium, 99% Ni.
GROUP 8	Stainless Steels Types 321 & 347, Cu-Ni, Monel, Inconel, Incoloy, Alloy 20
GROUP 9	Aluminum

Fabrication of alloy and non-ferrous piping is to be figured on the same basis as for corresponding operation on carbon steel materials, plus the percentage given below applicable to the carbon steel man hour schedules.

HANDLE AND ERECT FABRICATED SPOOL PIECES

PERCENT ADDITIVE

Nominal Pipe Size	MATERIAL CLASSIFICATION-GROUP NUMBERS AND PERCENTAGES								
	Group 1	Group 2	Group 3	Group 4	Group 5	Group 6	Group 7	Group 8	Group 9
2 or less	17.0	22.5	30.0	31.8	13.0	39.0	84.0	36.0	46.0
3	18.5	25.0	33.0	35.0	15.0	41.0	88.0	39.0	51.0
4	20.0	27.0	36.0	38.0	17.0	45.0	90.5	41.0	55.0
5	21.0	28.0	38.0	40.0	18.0	49.0	--	42.0	58.0
6	23.0	31.0	41.5	42.5	20.0	50.0	94.0	43.5	64.0
8	26.0	35.0	47.0	48.0	33.5	59.0	100.5	49.5	72.0
10	28.5	38.0	51.0	54.0	50.0	64.0	117.0	57.0	78.0
12	30.0	40.0	54.0	58.0	54.0	70.0	134.0	64.0	83.0
14	33.0	44.0	59.0	62.0	67.0	78.0	--	67.0	90.0
16	35.0	47.0	63.0	67.0	74.0	86.0	--	77.0	96.0
18	39.5	53.0	71.0	75.0	77.0	92.0	--	82.0	108.5
20	43.5	58.0	78.0	83.0	84.0	100.0	--	87.0	120.0
22	46.0	62.0	83.0	88.0	89.0	103.5	--	93.0	127.0
24	49.0	66.0	88.0	93.0	94.0	107.0	--	97.0	135.0
26	53.0	71.5	95.4	100.1	101.9	116.0	—	105.0	146.4
28	57.0	77.0	102.8	108.6	109.8	124.9	—	113.0	157.6
30	61.0	82.5	110.0	116.4	117.6	133.8	—	121.0	168.9
32	65.3	88.0	117.4	124.2	125.4	142.7	—	129.3	180.2
34	69.4	93.5	124.8	131.9	133.3	151.6	—	137.4	191.4
36	73.4	99.0	132.0	139.7	141.1	160.6	—	145.4	202.7
38	77.5	104.5	139.5	147.4	149.0	169.5	—	153.5	213.9
40	81.6	110.0	146.8	155.2	156.8	178.4	—	161.6	225.2
42	85.7	115.5	154.0	163.0	164.6	187.3	—	169.7	236.5

GROUP 1	ASTM A335-P1	.50% Moly	GROUP 4	Stainless Steels
	ASTM A335-P2	.50-.70% Chr.		Types 304, 309, 310, 316 (including "L" & "H" Grades)
	ASTM A335-P12	.85-1.10% Chr.		
	ASTM A335-P11	1.05-1.45% Chr.	GROUP 5	Copper, Brass, Everdur
	ASTM A335-P3	1.50-2.00% Chr.		
GROUP 2	ASTM A335-P3b	1.75-2.25% Chr.	GROUP 6	ASTM A-333-GR-1
	ASTM A335-P22	2.00-2.50% Chr.		ASTM A-333-GR4
	ASTM A335-P21	2.75-3.25% Chr.		ASTM A-333-GR9
	ASTM A335-P5,b,c	4.00-6.00% Chr.		
	ASTM A335-P7	6.00-8.00% Chr.	GROUP 7	Hastelloy, Titanium, 99% Ni.
	P9	8.00-10.00% Chr.	GROUP 8	Stainless Steels Types 321 & 347, Cu-Ni, Monel, Inconel, Incoloy, Alloy 20
	mes	10.00-15.00% Chr.		
	Gr. 3	3.50% Nickel	GROUP 9	Aluminum

Fabrication of alloy and non-ferrous piping is to be figured on the same basis as for corresponding operation on carbon steel materials, plus the percentage given below applicable to the carbon steel man hour schedules.

HANDLE AND ERECT STRAIGHT RUN PIPE (PERCENT ADDITIVE)

Nominal Pipe Size	MATERIAL CLASSIFICATION-GROUP NUMBERS AND PERCENTAGES								
	Group 1	Group 2	Group 3	Group 4	Group 5	Group 6	Group 7	Group 8	Group 9
2 or less	17.0	22.5	30.0	31.8	13.0	39.0	84.0	36.0	46.0
3	18.5	25.0	33.0	35.0	15.0	41.0	88.0	39.0	51.0
4	20.0	27.0	36.0	38.0	17.0	45.5	90.5	41.0	55.0
5	21.0	28.0	38.0	40.0	18.0	49.0	--	42.0	58.0
6	23.0	31.0	41.5	42.5	20.0	50.0	94.0	43.5	64.0
8	26.0	35.0	47.0	48.0	33.5	59.0	100.5	49.5	72.0
10	28.5	38.0	51.0	54.0	50.0	64.0	117.0	57.0	78.0
12	30.0	40.0	54.0	58.0	54.0	70.0	134.0	64.0	83.0
14	33.0	44.0	59.0	62.0	67.0	78.0	--	67.0	90.0
16	35.0	47.0	63.0	67.0	74.0	86.0	--	77.0	96.0
18	39.5	53.0	71.0	75.0	77.0	92.0	--	82.0	108.5
20	43.5	58.0	78.0	83.0	84.0	100.0	--	87.0	120.0
22	46.0	62.0	83.0	88.0	89.0	103.5	--	93.0	127.0
24	49.0	66.0	88.0	93.0	94.0	107.0	--	97.0	135.0
26	53.0	71.5	95.4	100.1	101.9	116.0	—	105.0	146.4
28	57.0	77.0	102.8	108.6	109.8	124.9	—	113.0	157.6
30	61.0	82.5	110.0	116.4	117.6	133.8	—	121.0	168.9
32	65.3	88.0	117.4	124.2	125.4	142.7	—	129.3	180.2
34	69.4	93.5	124.8	131.9	133.3	151.6	—	137.4	191.4
36	73.4	99.0	132.0	139.7	141.1	160.6	—	145.4	202.7
38	77.5	104.5	139.5	147.4	149.0	169.5	—	153.5	213.9
40	81.6	110.0	146.8	155.2	156.8	178.4	—	161.6	225.2
42	85.7	115.5	154.0	163.0	164.6	187.3	—	169.7	236.5

GROUP 1	ASTM A335-P1	.50% Moly
	ASTM A335-P2	.50-.70% Chr.
	ASTM A335-P12	.85-1.10% Chr.
	ASTM A335-P11	1.05-1.45% Chr.
	ASTM A335-P3	1.50-2.00% Chr.
GROUP 2	ASTM A335-P3b	1.75-2.25% Chr.
	ASTM A335-P22	2.00-2.50% Chr.
	ASTM A335-P21	2.75-3.25% Chr.
	ASTM A335-P5,b,c	4.00-6.00% Chr.
GROUP 3	ASTM A335-P7	6.00-8.00% Chr.
	ASTM A335-P9	8.00-10.00% Chr.
	Ferritic Chromes	10.00-15.00% Chr.
	ASTM A333 Gr. 3	3.50% Nickel

GROUP 4	Stainless Steels Types 304, 309, 310, 316 (including "L" & "H" Grades)
GROUP 5	Copper, Brass, Everdur
GROUP 6	ASTM A-333-GR-1 ASTM A-333-GR4 ASTM A-333-GR9
GROUP 7	Hastelloy, Titanium, 99% Ni.
GROUP 8	Stainless Steels Types 321 & 347, Cu-Ni, Monel, Inconel, Incoloy, Alloy 20
GROUP 9	Aluminum

Fabrication of alloy and non-ferrous piping is to be figured on the same basis as for corresponding operation on carbon steel materials, plus the percentage given below applicable to the carbon steel man hour schedules.

PIPE BENDS

PERCENT ADDITIVE

Nominal Pipe Size	MATERIAL CLASSIFICATION-GROUP NUMBERS AND PERCENTAGES								
	Group 1	Group 2	Group 3	Group 4	Group 5	Group 6	Group 7	Group 8	Group 9
2 or less	10.0	13.0	18.0	19.0	6.0	13.5	50.0	15.0	27.5
3	12.5	17.0	22.5	24.0	9.0	17.0	65.0	19.0	34.0
4	15.0	20.0	27.0	28.5	12.0	20.0	75.0	22.0	41.0
5	16.5	22.0	30.0	31.0	--	22.0	--	24.0	45.0
6	19.5	26.0	35.0	37.0	15.0	23.0	90.0	26.0	53.5
8	24.0	32.0	43.0	46.0	20.0	27.0	120.0	30.0	66.0
10	27.5	37.0	49.5	52.0	22.0	37.0	150.0	41.0	69.0
12	30.0	40.0	54.0	57.0	25.0	39.0	165.0	43.0	82.5
14	34.0	45.5	61.0	64.5	--	--	--	46.0	--
16	36.5	49.0	66.0	69.0	--	--	--	49.0	--
18	40.0	53.5	72.0	76.0	--	--	--	50.0	--
20	46.0	61.5	83.0	87.0	--	--	--	52.0	--
22	51.5	69.0	93.0	98.0	--	--	--	54.0	--
24	54.0	72.0	97.0	103.0	--	--	--	56.0	--

GROUP 1	ASTM A335-P1	.50% Moly	GROUP 4	Stainless Steels
	ASTM A335-P2	.50-.70% Chr.		Types 304, 309, 310, 316 (including "L" & "H" Grades)
	ASTM A335-P12	.85-1.10% Chr.		
	ASTM A335-P11	1.05-1.45% Chr.	GROUP 5	Copper, Brass, Everdur
	ASTM A335-P3	1.50-2.00% Chr.		
GROUP 2	ASTM A335-P3b	1.75-2.25% Chr.	GROUP 6	ASTM A-333-GR-1
	ASTM A335-P22	2.00-2.50% Chr.		ASTM A-333-GR4
	ASTM A335-P21	2.75-3.25% Chr.		ASTM A-333-GR9
	ASTM A335-P5,b,c	4.00-6.00% Chr.		
GROUP 3	ASTM A335-P7	6.00-8.00% Chr.	GROUP 7	Hastelloy, Titanium, 99% Ni.
	ASTM A335-P9	8.00-10.00% Chr.	GROUP 8	Stainless Steels Types 321 & 347, Cu-Ni, Monel, Inconel, Incoloy, Alloy 20
	Ferritic Chromes	10.00-15.00% Chr.		
	ASTM A333 Gr. 3	3.50% Nickel	GROUP 9	Aluminum

Fabrication of alloy and non-ferrous piping is to be figured on the same basis as for corresponding operation on carbon steel materials, plus the percentage given below applicable to the carbon steel man hour schedules.

ATTACHING FLANGES (PERCENT ADDITIVE)

Nominal Pipe Size	MATERIAL CLASSIFICATION-GROUP NUMBERS AND PERCENTAGES								
	Group 1	Group 2	Group 3	Group 4	Group 5	Group 6	Group 7	Group 8	Group 9
2 or less	25.0	33.5	45.0	47.5	20.0	58.0	125.0	54.0	69.0
3	27.5	37.0	49.5	52.0	23.0	61.0	132.0	58.0	76.0
4	30.0	40.0	54.0	57.0	25.0	68.0	135.0	61.0	82.5
5	31.5	42.0	57.0	60.0	27.5	73.0	--	63.0	87.0
6	34.5	46.0	62.0	63.5	30.0	75.0	140.0	65.0	95.0
8	39.0	52.0	70.0	72.0	50.0	87.5	150.0	74.0	107.0
10	42.5	57.0	76.5	81.0	75.0	95.0	175.0	85.0	117.0
12	45.0	60.0	81.0	86.0	80.0	104.0	200.0	95.0	124.0
14	49.0	65.5	88.0	93.0	100.0	117.0	--	100.0	135.0
16	52.5	70.0	94.5	100.0	110.0	128.0	--	115.0	144.0
18	59.0	79.0	106.0	112.0	115.0	138.0	--	123.0	162.0
20	65.0	87.0	117.0	123.5	125.0	149.0	--	130.0	179.0
22	69.0	92.9	124.0	131.0	133.0	154.5	--	139.0	190.0
24	73.0	98.0	131.0	139.0	140.0	160.0	--	145.0	201.0
26	79.0	106.0	142.0	150.4	151.6	173.4	—	157.0	217.9
28	85.0	114.2	152.9	162.0	163.2	186.8	—	169.0	234.6
30	91.2	122.4	163.8	173.7	174.9	200.0	—	181.2	251.4
32	97.3	130.6	174.7	185.3	186.6	213.4	—	193.3	268.2
34	103.4	138.7	185.6	196.9	198.2	226.8	—	205.4	284.9
36	109.4	146.9	196.6	208.4	209.9	240.0	—	217.4	301.7
38	115.5	155.0	207.5	220.0	221.5	253.5	—	229.5	318.4
40	121.6	163.2	218.4	231.6	233.2	266.8	—	241.6	335.2
42	127.7	171.4	229.3	243.2	244.9	280.0	—	253.7	352.0

GROUP 1	ASTM A335-P1	.50% Moly
	ASTM A335-P2	.50-.70% Chr.
	ASTM A335-P12	.85-1.10% Chr.
	ASTM A335-P11	1.05-1.45% Chr.
	ASTM A335-P3	1.50-2.00% Chr.

GROUP 2	ASTM A335-P3b	1.75-2.25% Chr.
	ASTM A335-P22	2.00-2.50% Chr.
	ASTM A335-P21	2.75-3.25% Chr.
	ASTM A335-P5,b,c	4.00-6.00% Chr.

GROUP 3	ASTM A335-P7	6.00-8.00% Chr.
	ASTM A335-P9	8.00-10.00% Chr.
	Ferritic Chromes	10.00-15.00% Chr.
	ASTM A333 Gr. 3	3.50% Nickel

GROUP 4	Stainless Steels Types 304, 309, 310, 316 (including "L" & "H" Grades)
GROUP 5	Copper, Brass, Everdur
GROUP 6	ASTM A-333-GR-1 ASTM A-333-GR4 ASTM A-333-GR9
GROUP 7	Hastelloy, Titanium, 99% Ni.
GROUP 8	Stainless Steels Types 321 & 347, Cu-Ni, Monel, Inconel, Incoloy, Alloy 20
GROUP 9	Aluminum

Fabrication of alloy and non-ferrous piping is to be figured on the same basis as for corresponding operation on carbon steel materials, plus the percentage given below applicable to the carbon steel man hour schedules.

MAKE-ONS THROUGH 12-IN.
HANDLE VALVES THROUGH 42-IN.

PERCENT ADDITIVE

Nominal Size Inches	MATERIAL CLASSIFICATION-GROUP NUMBERS AND PERCENTAGES								
	Group 1	Group 2	Group 3	Group 4	Group 5	Group 6	Group 7	Group 8	Group 9
1/4	9.0	11.0	16.0	17.0	7.0	21.0	45.0	20.0	25.0
3/8	10.0	12.5	18.0	19.0	8.0	23.0	50.0	22.0	28.0
1/2	11.0	14.0	20.0	21.0	9.0	26.0	56.0	24.0	31.0
3/4	12.0	16.0	22.5	23.0	10.0	29.0	62.0	27.0	34.0
1	13.5	18.0	25.0	26.0	11.0	32.0	69.0	30.0	38.0
1-1/4	15.0	20.0	27.5	29.0	12.0	36.0	76.5	33.0	42.0
1-1/2	17.0	22.5	30.5	32.0	13.5	40.0	85.0	37.0	47.0
2	19.0	25.0	34.0	36.0	15.0	44.0	94.0	41.0	52.0
2-1/2	20.5	27.0	35.5	37.5	16.0	45.5	97.0	42.5	55.0
3	21.0	28.0	37.0	39.0	17.0	46.0	99.0	44.0	57.0
3-1/2	22.5	29.0	39.5	42.0	18.0	49.5	100.0	45.0	60.5
4	23.0	30.0	41.0	43.0	19.0	51.0	101.0	46.0	62.0
6	24.2	31.6	43.3	45.5	19.4	55.3	105.0	51.4	66.0
8	32.2	42.2	57.8	60.7	25.9	73.8	140.0	68.6	88.0
10	40.3	52.7	72.2	75.9	32.4	92.2	175.0	85.7	110.0
12	48.4	63.2	86.6	91.1	38.9	110.6	210.0	102.8	132.0
14	56.4	73.8	101.0	106.3	45.4	129.0	—	120.0	154.0
16	64.5	84.3	115.5	121.4	51.8	147.5	—	137.0	176.0
18	72.5	94.9	130.0	136.6	58.3	166.0	—	154.3	198.0
20	80.1	105.4	158.8	151.8	64.8	184.4	—	171.4	220.0
22	88.7	115.9	173.3	167.0	71.3	202.8	—	188.5	242.0
24	96.7	126.5	187.7	182.2	77.8	221.3	—	205.7	264.0
26	104.8	137.0	187.7	197.3	84.2	239.7	—	222.8	286.0
28	112.8	147.6	202.2	212.5	90.7	258.2	—	240.0	308.0
30	120.9	158.0	216.6	227.7	97.2	276.6	—	257.0	330.0
32	129.0	168.6	231.0	242.9	103.7	295.0	—	274.2	352.0
34	137.0	179.2	245.5	258.1	110.2	313.5	—	291.4	374.0
36	145.0	189.7	259.9	273.2	116.6	331.9	—	308.5	396.0
38	153.0	200.3	274.4	288.4	123.0	350.4	—	325.6	418.0
40	161.2	210.8	288.8	303.6	129.6	368.8	—	342.8	440.0
42	169.3	221.3	303.2	318.8	136.0	387.2	—	359.9	462.0

(table continued on next page)

MAKE-ONS THROUGH 12-IN.
HANDLE VALVES THROUGH 42-IN.

(CONTINUED)

GROUP 1	ASTM A335-P1	.50% Moly	**GROUP 4**	**Stainless Steels** Types 304, 309, 310, 316 (including "L" & "H" Grades)
	ASTM A335-P2	.50-.70% Chr.		
	ASTM A335-P12	.85-1.10% Chr.		
	ASTM A335-P11	1.05-1.45% Chr.	**GROUP 5**	Copper, Brass, Everdur
	ASTM A335-P3	1.50-2.00% Chr.		
GROUP 2	ASTM A335-P3b	1.75-2.25% Chr.	**GROUP 6**	ASTM A-333-GR-1 ASTM A-333-GR4 ASTM A-333-GR9
	ASTM A335-P22	2.00-2.50% Chr.		
	ASTM A335-P21	2.75-3.25% Chr.		
	ASTM A335-P5,b,c	4.00-6.00% Chr.		
GROUP 3	ASTM A335-P7	6.00-8.00% Chr.	**GROUP 7**	Hastelloy, Titanium, 99% Ni.
	ASTM A335-P9	8.00-10.00% Chr.	**GROUP 8**	Stainless Steels Types 321 & 347, Cu-Ni, Monel, Inconel, Incoloy, Alloy 20
	Ferritic Chromes	10.00-15.00% Chr.		
	ASTM A333 Gr. 3	3.50% Nickel	**GROUP 9**	Aluminum

Fabrication of alloy and non-ferrous piping is to be figured on the same basis as for corresponding operation on carbon steel materials, plus the percentage given below applicable to the carbon steel man hour schedules.

FIELD ERECTION BOLT-UPS (PERCENT ADDITIVE)

Nominal Pipe Size	MATERIAL CLASSIFICATION-GROUP NUMBERS AND PERCENTAGES								
	Group 1	Group 2	Group 3	Group 4	Group 5	Group 6	Group 7	Group 8	Group 9
2 or less	24.0	32.0	43.0	45.0	19.0	55.0	119.0	51.0	66.0
3	26.0	35.0	47.0	49.0	22.0	58.0	125.0	55.0	72.0
4	28.5	38.0	51.0	54.0	24.0	65.0	128.0	58.0	78.0
5	30.0	40.0	54.0	57.0	26.0	69.0	--	60.0	83.0
6	33.0	44.0	59.0	60.0	28.5	71.0	133.0	62.0	90.0
8	37.0	49.0	66.5	68.0	47.5	83.0	143.0	70.0	102.0
10	40.0	54.0	73.0	77.0	71.0	90.0	166.0	81.0	111.0
12	43.0	57.0	77.0	82.0	76.0	99.0	190.0	90.0	118.0
14	47.0	62.0	84.0	88.0	95.0	111.0	--	95.0	128.0
16	50.0	66.5	90.0	95.0	105.0	122.0	--	109.0	137.0
18	56.0	75.0	101.0	106.0	109.0	131.0	--	117.0	154.0
20	62.0	83.0	111.0	117.0	119.0	142.0	--	124.0	170.0
22	65.0	88.0	118.0	124.0	126.0	147.0	--	132.0	181.0
24	69.0	93.0	124.0	132.0	133.0	152.0	--	138.0	191.0
26	74.9	100.1	134.4	143.0	144.0	164.6	—	149.5	207.0
28	80.6	108.6	144.8	154.0	155.0	177.2	—	161.0	222.9
30	86.4	116.4	155.0	165.0	166.2	189.9	—	172.5	238.8
32	92.2	124.2	165.4	176.0	177.3	202.6	—	184.0	254.7
34	97.9	131.9	175.8	187.0	188.4	215.2	—	195.5	270.6
36	103.7	139.7	186.1	198.0	199.4	227.9	—	207.0	286.6
38	109.4	147.4	196.5	209.0	210.5	240.5	—	218.5	302.5
40	115.2	155.2	206.8	220.0	221.6	253.2	—	230.0	318.4
42	121.0	163.0	217.0	231.0	232.7	265.9	—	241.5	334.3

GROUP 1	ASTM A335-P1	.50% Moly	GROUP 4	Stainless Steels Types 304, 309, 310, 316 (including "L" & "H" Grades)
	ASTM A335-P2	.50-.70% Chr.		
	ASTM A335-P12	.85-1.10% Chr.		
	ASTM A335-P11	1.05-1.45% Chr.	GROUP 5	Copper, Brass, Everdur
	ASTM A335-P3	1.50-2.00% Chr.		
GROUP 2	ASTM A335-P3b	1.75-2.25% Chr.	GROUP 6	ASTM A-333-GR-1 ASTM A-333-GR4 ASTM A-333-GR9
	ASTM A335-P22	2.00-2.50% Chr.		
	ASTM A335-P21	2.75-3.25% Chr.		
	ASTM A335-P5,b,c	4.00-6.00% Chr.		
GROUP 3	ASTM A335-P7	6.00-8.00% Chr.	GROUP 7	Hastelloy, Titanium, 99% Ni.
	ASTM A335-P9	8.00-10.00% Chr.	GROUP 8	Stainless Steels Types 321 & 347, Cu-Ni, Monel, Inconel, Incoloy, Alloy 20
	Ferritic Chromes	10.00-15.00% Chr.		
	ASTM A333 Gr. 3	3.50% Nickel	GROUP 9	Aluminum

Fabrication of alloy and non-ferrous piping is to be figured on the same basis as for corresponding operation on carbon steel materials, plus the percentage given below applicable to the carbon steel man hour schedules.

ALL WELDED FABRICATION (PERCENT ADDITIVE)

Nominal Pipe Size	MATERIAL CLASSIFICATION-GROUP NUMBERS AND PERCENTAGES								
	Group 1	Group 2	Group 3	Group 4	Group 5	Group 6	Group 7	Group 8	Group 9
2 or less	25.0	33.5	45.0	47.5	20.0	58.0	125.0	54.0	69.0
3	27.5	37.0	49.5	52.0	23.0	61.0	132.0	58.0	76.0
4	30.0	40.0	54.0	57.0	25.0	68.0	135.0	61.0	82.5
5	31.5	42.0	57.0	60.0	27.5	73.0	--	63.0	87.0
6	34.5	46.0	62.0	63.5	30.0	75.0	140.0	65.0	95.0
8	39.0	52.0	70.0	72.0	50.0	87.5	150.0	74.0	107.0
10	42.5	57.0	76.5	81.0	75.0	95.0	175.0	85.0	117.0
12	45.0	60.0	81.0	86.0	80.0	104.0	200.0	95.0	124.0
14	49.0	65.5	88.0	93.0	100.0	117.0	--	100.0	135.0
16	52.5	70.0	94.5	100.0	110.0	128.0	--	115.0	144.0
18	59.0	79.0	106.0	112.0	115.0	138.0	—	123.0	162.0
20	65.0	87.0	117.0	123.5	125.0	149.0	--	130.0	179.0
22	69.0	92.5	124.0	131.0	133.0	154.5	—	139.0	190.0
24	73.0	98.0	131.0	139.0	140.0	160.0	--	145.0	201.0
26	79.0	106.0	142.0	150.5	151.6	173.4	—	157.0	217.9
28	85.0	114.2	152.9	162.0	163.2	186.8	—	169.0	234.6
30	91.2	122.4	163.8	173.7	174.9	200.0	—	181.2	251.4
32	97.3	130.6	174.7	185.3	186.6	213.4	—	193.3	268.2
34	103.4	138.7	185.6	196.9	198.2	226.8	—	205.4	284.9
36	109.4	146.9	196.6	208.4	209.9	240.0	—	217.4	301.7
38	115.5	155.0	207.5	220.0	221.5	253.5	—	229.5	318.4
40	121.6	163.2	218.4	231.6	233.2	266.8	—	241.6	335.2
42	127.7	171.4	229.3	243.2	244.9	280.0	—	253.7	352.0

GROUP 1	ASTM A335-P1	.50% Moly	**GROUP 4**	Stainless Steels	
	ASTM A335-P2	.50-.70% Chr.		Types 304, 309, 310, 316 (including "L" & "H" Grades)	
	ASTM A335-P12	.85-1.10% Chr.			
	ASTM A335-P11	1.05-1.45% Chr.	**GROUP 5**	Copper, Brass, Everdur	
	ASTM A335-P3	1.50-2.00% Chr.			
GROUP 2	ASTM A335-P3b	1.75-2.25% Chr.	**GROUP 6**	ASTM A-333-GR-1	
	ASTM A335-P22	2.00-2.50% Chr.		ASTM A-333-GR4	
	ASTM A335-P21	2.75-3.25% Chr.		ASTM A-333-GR9	
	ASTM A335-P5,b,c	4.00-6.00% Chr.			
GROUP 3	ASTM A335-P7	6.00-8.00% Chr.	**GROUP 7**	Hastelloy, Titanium, 99% Ni.	
	ASTM A335-P9	8.00-10.00% Chr.	**GROUP 8**	Stainless Steels Types 321 & 347, Cu-Ni, Monel, Inconel, Incoloy, Alloy 20	
	Ferritic Chromes	10.00-15.00% Chr.			
	ASTM A333 Gr. 3	3.50% Nickel	**GROUP 9**	Aluminum	

Fabrication of alloy and non-ferrous piping is to be figured on the same basis as for corresponding operation on carbon steel materials, plus the percentage given below applicable to the carbon steel man hour schedules.

FLAME CUTTING OR BEVELING (PERCENT ADDITIVE)

Nominal Pipe Size	MTR'L. CLASSIFICATION		
	Group 1	Group 2	Group 6
2 or less	17.0	22.5	39.0
3	18.5	25.0	41.0
4	20.0	27.0	45.5
5	21.0	28.0	49.0
6	23.0	31.0	50.0
8	26.0	35.0	59.0
10	28.5	38.0	64.0
12	30.0	40.0	70.0
14	33.0	44.0	78.0
16	35.0	47.0	86.0
18	39.5	53.0	92.0
20	43.5	58.0	100.0
22	46.0	62.0	103.5
24	49.0	66.0	107.0
26	53.0	71.5	116.0
28	57.0	77.0	124.9
30	61.0	82.5	133.8
32	65.3	88.0	142.7
34	69.4	93.5	151.6
36	73.4	99.0	160.6
38	77.5	104.5	169.5
40	81.6	110.0	178.4
42	85.7	115.5	187.3

Material in Groups 1, 2 and 6 only will flame cut. All others are to be machine cut. See respective percentage tables for others.

GROUP 1	ASTM A335-P1	.50% Moly	GROUP 4	Stainless Steels Types 304, 309, 310, 316 (including "L" & "H" Grades)
	ASTM A335-P2	.50-.70% Chr.		
	ASTM A335-P12	.85-1.10% Chr.		
	ASTM A335-P11	1.05-1.45% Chr.	GROUP 5	Copper, Brass, Everdur
	ASTM A335-P3	1.50-2.00% Chr.		
GROUP 2	ASTM A335-P3b	1.75-2.25% Chr.	GROUP 6	ASTM A-333-GR-1 ASTM A-333-GR4 ASTM A-333-GR9
	ASTM A335-P22	2.00-2.50% Chr.		
	ASTM A335-P21	2.75-3.25% Chr.		
	ASTM A335-P5,b,c	4.00-6.00% Chr.		
GROUP 3	ASTM A335-P7	6.00-8.00% Chr.	GROUP 7	Hastelloy, Titanium, 99% Ni.
	ASTM A335-P9	8.00-10.00% Chr.	GROUP 8	Stainless Steels Types 321 & 347, Cu-Ni, Monel, Inconel, Incoloy, Alloy 20
	Ferritic Chromes	10.00-15.00% Chr.		
	ASTM A333 Gr. 3	3.50% Nickel	GROUP 9	Aluminum

Fabrication of alloy and non-ferrous piping is to be figured on the same basis as for corresponding operation on carbon steel materials, plus the percentage given below applicable to the carbon steel man hour schedules.

MACHINE CUTTING AND BEVELING PIPE (PERCENT ADDITIVE)

Nominal Pipe Size	MATERIAL CLASSIFICATION-GROUP NUMBERS AND PERCENTAGES								
	Group 1	Group 2	Group 3	Group 4	Group 5	Group 6	Group 7	Group 8	Group 9
2 or less	17.0	22.5	30.0	31.8	13.0	39.0	84.0	36.0	46.0
3	18.5	25.0	33.0	35.0	15.0	41.0	88.0	39.0	51.0
4	20.0	27.0	36.0	38.0	17.0	45.5	90.5	41.0	55.0
5	21.0	28.0	38.0	40.0	18.0	49.0	--	42.0	58.0
6	23.0	31.0	41.5	42.5	20.0	50.0	94.0	43.5	64.0
8	26.0	35.0	47.0	48.0	33.5	59.0	100.5	49.5	72.0
10	28.5	38.0	51.0	54.0	50.0	64.0	117.0	57.0	78.0
12	30.0	40.0	54.0	58.0	54.0	70.0	134.0	64.0	83.0
14	33.0	44.0	59.0	62.0	67.0	78.0	--	67.0	90.0
16	35.0	47.0	63.0	67.0	74.0	86.0	--	77.0	96.0
18	39.5	53.0	71.0	75.0	77.0	92.0	--	82.0	108.5
20	43.5	58.0	78.0	83.0	84.0	100.0	--	87.0	120.0
22	46.0	62.0	83.0	88.0	89.0	103.5	--	93.0	127.0
24	49.0	66.0	88.0	93.0	94.0	107.0	--	97.0	135.0
26	53.0	71.5	95.4	100.1	101.9	116.0	—	105.0	146.4
28	57.0	77.0	102.8	108.6	109.8	124.9	—	113.0	157.6
30	61.0	82.5	110.0	116.4	117.6	133.8	—	121.0	168.9
32	65.3	88.0	117.4	124.2	125.4	142.7	—	129.3	180.2
34	69.4	93.5	124.8	131.9	133.3	151.6	—	137.4	191.4
36	73.4	99.0	132.0	139.7	141.1	160.6	—	145.4	202.7
38	77.5	104.5	139.5	147.4	149.0	169.5	—	153.5	213.9
40	81.6	110.0	146.8	155.2	156.8	178.4	—	161.6	225.2
42	85.7	115.5	154.0	163.0	164.6	187.3	—	169.7	236.5

GROUP 1	ASTM A335-P1	.50% Moly	GROUP 4	Stainless Steels
	ASTM A335-P2	.50-.70% Chr.		Types 304, 309, 310, 316 (including "L" & "H" Grades)
	ASTM A335-P12	.85-1.10% Chr.		
	ASTM A335-P11	1.05-1.45% Chr.	GROUP 5	Copper, Brass, Everdur
	ASTM A335-P3	1.50-2.00% Chr.		
GROUP 2	ASTM A335-P3b	1.75-2.25% Chr.	GROUP 6	ASTM A-333-GR-1
	ASTM A335-P22	2.00-2.50% Chr.		ASTM A-333-GR4
	ASTM A335-P21	2.75-3.25% Chr.		ASTM A-333-GR9
	ASTM A335-P5,b,c	4.00-6.00% Chr.		
GROUP 3	ASTM A335-P7	6.00-8.00% Chr.	GROUP 7	Hastelloy, Titanium, 99% Ni.
	ASTM A335-P9	8.00-10.00% Chr.	GROUP 8	Stainless Steels Types 321 & 347, Cu-Ni, Monel, Inconel, Incoloy, Alloy 20
	Ferritic Chromes	10.00-15.00% Chr.		
	ASTM A333 Gr. 3	3.50% Nickel	GROUP 9	Aluminum

Fabrication of alloy and non-ferrous piping is to be figured on the same basis as for corresponding operation on carbon steel materials, plus the percentage given below applicable to the carbon steel man hour schedules.

THREADING PIPE

PERCENT ADDITIVE

Nominal Size Inches	MATERIAL CLASSIFICATION-GROUP NUMBERS AND PERCENTAGE								
	Group 1	Group 2	Group 3	Group 4	Group 5	Group 6	Group 7	Group 8	Group 9
2 or less	20.0	27.0	36.0	38.0	16.0	46.0	100.0	43.0	55.0
3	22.0	30.0	40.0	42.0	18.0	49.0	106.0	46.0	61.0
4	24.0	32.0	43.0	46.0	20.0	54.0	108.0	49.0	66.0
5	25.0	34.0	46.0	48.0	22.0	58.0	--	50.0	70.0
6	28.0	37.0	50.0	51.0	24.0	60.0	112.0	52.0	76.0
8	31.0	42.0	56.0	58.0	40.0	70.0	120.0	59.0	86.0
10	34.0	46.0	61.0	65.0	60.0	76.0	140.0	68.0	94.0
12	36.0	48.0	65.0	69.0	64.0	83.0	160.0	76.0	99.0

GROUP 1	ASTM A335-P1	.50% Moly	**GROUP 4**	Stainless Steels Types 304, 309, 310, 316 (including "L" & "H" Grades)	
	ASTM A335-P2	.50-.70% Chr.			
	ASTM A335-P12	.85-1.10% Chr.			
	ASTM A335-P11	1.05-1.45% Chr.	**GROUP 5**	Copper, Brass, Everdur	
	ASTM A335-P3	1.50-2.00% Chr.			
GROUP 2	ASTM A335-P3b	1.75-2.25% Chr.	**GROUP 6**	ASTM A-333-GR-1 ASTM A-333-GR4 ASTM A-333-GR9	
	ASTM A335-P22	2.00-2.50% Chr.			
	ASTM A335-P21	2.75-3.25% Chr.			
	ASTM A335-P5,b,c	4.00-6.00% Chr.			
GROUP 3	ASTM A335-P7	6.00-8.00% Chr.	**GROUP 7**	Hastelloy, Titanium, 99% Ni.	
	ASTM A335-P9	8.00-10.00% Chr.	**GROUP 8**	Stainless Steels Types 321 & 347, Cu-Ni, Monel, Inconel, Incoloy, Alloy 20	
	Ferritic Chromes	10.00-15.00% Chr.	**GROUP 9**	Aluminum	
	ASTM A333 Gr. 3	3.50% Nickel			

Fabrication of alloy and non-ferrous piping is to be figured on the same basis as for corresponding operation on carbon steel materials, plus the percentage given below applicable to the carbon steel man hour schedules.

WELDED ATTACHMENTS AND DRILLING HOLES IN WELDED ATTACHMENTS

PERCENT ADDITIVE

Thickness of Plate, Angle Etc., Inches	MATERIAL CLASSIFICATION-GROUP NUMBERS AND PERCENTAGES								
	Group 1	Group 2	Group 3	Group 4	Group 5	Group 6	Group 7	Group 8	Group 9
1/2 or less	21.0	29.0	40.0	43.0	16.0	53.0	98.0	49.0	62.0
3/4	21.5	29.5	40.5	43.5	16.5	53.5	102.0	49.5	62.5
1	22.0	30.0	41.0	44.0	17.0	54.0	106.0	50.0	63.0
1-1/4	22.5	30.5	41.5	44.5	17.5	54.5	110.0	50.5	63.5
1-1/2	23.0	31.0	42.0	45.0	18.0	55.0	115.0	51.0	64.0
1-3/4	24.0	32.0	43.0	46.0	19.0	56.0	120.0	52.0	66.0
2	25.0	33.5	45.0	47.5	20.0	58.0	125.0	54.0	69.0
2-1/2	26.5	35.0	47.0	50.5	22.0	59.0	129.0	56.0	73.5
3	27.5	37.0	49.5	52.0	23.0	61.0	132.0	58.0	76.0
3-1/2	29.0	39.0	52.5	55.0	24.5	65.0	134.0	59.5	79.0
4	30.0	40.0	54.0	57.0	25.0	68.0	135.0	61.0	82.5

GROUP 1	ASTM A335-P1	.50% Moly	GROUP 4	Stainless Steels Types 304, 309, 310, 316 (including "L" & "H" Grades)
	ASTM A335-P2	.50-.70% Chr.		
	ASTM A335-P12	.85-1.10% Chr.		
	ASTM A335-P11	1.05-1.45% Chr.	GROUP 5	Copper, Brass, Everdur
	ASTM A335-P3	1.50-2.00% Chr.		
GROUP 2	ASTM A335-P3b	1.75-2.25% Chr.	GROUP 6	ASTM A-333-GR-1 ASTM A-333-GR4 ASTM A-333-GR9
	ASTM A335-P22	2.00-2.50% Chr.		
	ASTM A335-P21	2.75-3.25% Chr.		
	ASTM A335-P5,b,c	4.00-6.00% Chr.		
GROUP 3	ASTM A335-P7	6.00-8.00% Chr.	GROUP 7	Hastelloy, Titanium, 99% Ni.
	ASTM A335-P9	8.00-10.00% Chr.	GROUP 8	Stainless Steels Types 321 & 347, Cu-Ni, Monel, Inconel, Incoloy, Alloy 20
	Ferritic Chromes	10.00-15.00% Chr.		
	ASTM A333 Gr. 3	3.50% Nickel	GROUP 9	Aluminum

Fabrication of alloy and non-ferrous piping is to be figured on the same basis as for corresponding operation on carbon steel materials, plus the percentage given below applicable to the carbon steel man hour schedules.

LOCAL STRESS RELIEVING (PERCENT ADDITIVE)

Nominal Pipe Size	MATERIAL CLASSIFICATION-GROUP NUMBERS AND PERCENTAGES								
	Group 1	Group 2	Group 3	Group 4	Group 5	Group 6	Group 7	Group 8	Group 9
2 or less	17.0	22.5	30.0	31.8	13.0	39.0	84.0	36.0	46.0
3	18.5	25.0	33.0	35.0	15.0	41.0	88.0	39.0	51.0
4	20.0	27.0	36.0	38.0	17.0	45.5	90.5	41.0	55.0
5	21.0	28.0	38.0	40.0	18.0	49.0	--	42.0	58.0
6	23.0	31.0	41.5	42.5	20.0	50.0	94.0	43.5	64.0
8	26.0	35.0	47.0	48.0	33.5	59.0	100.5	49.5	72.0
10	28.5	38.0	51.0	54.0	50.0	64.0	117.0	57.0	78.0
12	30.0	40.0	54.0	58.0	54.0	70.0	134.0	64.0	83.0
14	33.0	44.0	59.0	62.0	67.0	78.0	--	67.0	90.0
16	35.0	47.0	63.0	67.0	74.0	86.0	--	77.0	96.0
18	39.5	53.0	71.0	75.0	77.0	92.0	--	82.0	108.5
20	43.5	58.0	78.0	83.0	84.0	100.0	--	87.0	120.0
22	46.0	62.0	83.0	88.0	89.0	103.5	--	93.0	127.0
24	49.0	66.0	88.0	93.0	94.0	107.0	--	97.0	135.0
26	53.0	71.5	95.4	100.1	101.9	116.0	—	105.0	146.4
28	57.0	77.0	102.8	108.6	109.8	124.9	—	113.0	157.6
30	61.0	82.5	110.0	116.4	117.6	133.8	—	121.0	168.9
32	65.3	88.0	117.4	124.2	125.4	142.7	—	129.3	180.2
34	69.4	93.5	124.8	131.9	133.3	151.6	—	137.4	191.4
36	73.4	99.0	132.0	139.7	141.1	160.6	—	145.4	202.7
38	77.5	104.5	139.5	147.4	149.0	169.5	—	153.5	213.9
40	81.6	110.0	146.8	155.2	156.8	178.4	—	161.6	225.2
42	85.7	115.5	154.0	163.0	164.6	187.3	—	169.7	236.5

GROUP 1	ASTM A335-P1	.50% Moly	GROUP 4	Stainless Steels Types 304, 309, 310, 316 (including "L" & "H" Grades)
	ASTM A335-P2	.50-.70% Chr.		
	ASTM A335-P12	.85-1.10% Chr.		
	ASTM A335-P11	1.05-1.45% Chr.	GROUP 5	Copper, Brass, Everdur
	ASTM A335-P3	1.50-2.00% Chr.		
GROUP 2	ASTM A335-P3b	1.75-2.25% Chr.	GROUP 6	ASTM A-333-GR-1 ASTM A-333-GR4 ASTM A-333-GR9
	ASTM A335-P22	2.00-2.50% Chr.		
	ASTM A335-P21	2.75-3.25% Chr.		
	ASTM A335-P5,b,c	4.00-6.00% Chr.		
GROUP 3	ASTM A335-P7	6.00-8.00% Chr.	GROUP 7	Hastelloy, Titanium, 99% Ni.
	ASTM A335-P9	8.00-10.00% Chr.	GROUP 8	Stainless Steels Types 321 & 347, Cu-Ni, Monel, Inconel, Incoloy, Alloy 20
	Ferritic Chromes	10.00-15.00% Chr.		
	ASTM A333 Gr. 3	3.50% Nickel	GROUP 9	Aluminum

Fabrication of alloy and non-ferrous piping is to be figured on the same basis as for corresponding operation on carbon steel materials, plus the percentage given below applicable to the carbon steel man hour schedules.

RADIOGRAPHIC INSPECTION (PERCENT ADDITIVE)

Nominal Pipe Size	MATERIAL CLASSIFICATION-GROUP NUMBERS AND PERCENTAGES								
	Group 1	Group 2	Group 3	Group 4	Group 5	Group 6	Group 7	Group 8	Group 9
2 or less	5.0	7.0	15.0	9.5	4.0	11.5	42.5	18.0	34.5
3	5.5	7.0	17.0	10.0	4.5	12.0	45.0	20.0	38.0
4	6.0	8.0	18.0	11.5	5.0	13.5	46.0	21.0	41.0
5	6.5	8.5	19.0	12.0	5.5	14.5	--	21.5	43.5
6	7.0	9.0	21.0	13.0	6.0	15.0	47.5	22.0	47.5
8	8.0	10.0	24.0	14.0	10.0	17.5	51.0	25.0	53.5
10	9.0	11.0	26.0	16.0	15.0	19.0	59.5	29.0	58.5
12	9.5	12.0	27.5	17.0	16.0	21.0	68.0	32.0	62.0
14	10.0	13.0	30.0	19.0	20.0	23.0	--	34.0	67.5
16	10.5	14.0	32.0	20.0	22.0	25.5	--	39.0	72.0
18	12.0	16.0	36.0	22.0	23.0	27.5	--	42.0	81.0
20	13.0	17.0	40.0	25.0	25.0	30.0	--	44.0	89.5
22	14.0	18.5	42.0	26.0	27.0	31.0	--	47.0	95.0
24	15.0	20.0	44.5	28.0	28.0	32.0	--	49.0	101.0
26	16.4	21.6	48.1	30.4	30.4	34.6	—	53.0	109.2
28	17.6	23.2	51.8	32.8	32.8	37.2	—	57.0	117.6
30	18.9	24.9	55.5	35.1	35.1	39.9	—	61.0	126.0
32	20.2	26.6	59.2	37.4	37.4	42.6	—	65.3	134.4
34	21.4	28.2	62.9	39.8	39.8	45.2	—	69.4	142.8
36	22.7	29.9	66.5	42.1	42.1	47.9	—	73.4	151.2
38	23.9	31.5	70.3	44.5	44.5	50.5	—	77.5	159.6
40	25.2	33.2	74.0	46.8	46.8	53.2	—	81.6	168.0
42	26.5	34.9	77.7	49.1	49.1	55.9	—	85.7	176.4

GROUP 1	ASTM A335-P1	.50% Moly	**GROUP 4**	Stainless Steels Types 304, 309, 310, 316 (including "L" & "H" Grades)
	ASTM A335-P2	.50-.70% Chr.		
	ASTM A335-P12	.85-1.10% Chr.		
	ASTM A335-P11	1.05-1.45% Chr.	**GROUP 5**	Copper, Brass, Everdur
	ASTM A335-P3	1.50-2.00% Chr.		
GROUP 2	ASTM A335-P3b	1.75-2.25% Chr.	**GROUP 6**	ASTM A-333-GR-1 ASTM A-333-GR4 ASTM A-333-GR9
	ASTM A335-P22	2.00-2.50% Chr.		
	ASTM A335-P21	2.75-3.25% Chr.		
	ASTM A335-P5,b,c	4.00-6.00% Chr.		
GROUP 3	ASTM A335-P7	6.00-8.00% Chr.	**GROUP 7**	Hastelloy, Titanium, 99% Ni.
	ASTM A335-P9	8.00-10.00% Chr.	**GROUP 8**	Stainless Steels Types 321 & 347, Cu-Ni, Monel, Inconel, Incoloy, Alloy 20
	Ferritic Chromes	10.00-15.00% Chr.		
	ASTM A333 Gr. 3	3.50% Nickel	**GROUP 9**	Aluminum

Fabrication of alloy and non-ferrous piping is to be figured on the same basis as for corresponding operation on carbon steel materials, plus the percentage given below applicable to the carbon steel man hour schedules.

MAGNETIC OR DYE PENETRANT INSPECTION (PERCENT ADDITIVE)

| Nominal Pipe Size | MATERIAL CLASSIFICATION—GROUP NUMBERS AND PERCENTAGES | | | | | | | | |
	Group 1	Group 2	Group 3	Group 4	Group 5	Group 6	Group 7	Group 8	Group 9
2 or less	5.0	7.0	9.0	9.5	4.0	11.5	25.0	11.0	14.0
3	5.5	7.0	10.0	10.0	4.5	12.0	26.0	11.5	15.0
4	6.0	8.0	11.0	11.5	5.0	13.5	27.0	12.0	16.5
5	6.5	8.5	11.5	12.0	5.5	14.5	--	12.5	17.0
6	7.0	9.0	12.0	13.0	6.0	15.0	28.0	13.0	19.0
8	8.0	10.0	14.0	14.0	10.0	17.5	30.0	15.0	21.0
10	9.0	11.0	15.0	16.0	15.0	19.0	35.0	17.0	23.0
12	9.5	12.0	16.0	17.0	16.0	21.0	40.0	19.0	25.0
14	10.0	13.0	17.5	19.0	20.0	23.0	--	20.0	27.0
16	10.5	14.0	19.0	20.0	22.0	25.5	--	23.0	29.0
18	12.0	16.0	21.0	22.0	23.0	27.5	--	24.5	32.0
20	13.0	17.0	23.0	25.0	25.0	30.0	--	26.0	36.0
22	14.0	18.5	25.0	26.0	27.0	31.0	--	28.0	38.0
24	15.0	20.0	26.0	28.0	28.0	32.0	--	29.0	40.0
26	16.4	21.6	28.1	30.4	30.4	34.6	—	31.2	43.4
28	17.6	23.2	30.2	32.8	32.8	37.2	—	33.6	46.8
30	18.9	24.9	32.4	35.1	35.1	39.9	—	36.0	50.1
32	20.2	26.6	34.6	37.4	37.4	42.6	—	38.4	53.4
34	21.4	28.2	36.7	39.8	39.8	45.2	—	40.8	56.8
36	22.7	29.9	38.9	42.1	42.1	47.9	—	43.2	60.1
38	23.9	31.5	41.0	44.5	44.5	50.5	—	45.6	63.5
40	25.2	33.2	43.2	46.8	46.8	53.2	—	48.0	66.8
42	26.5	34.9	45.4	49.1	49.1	55.9	—	50.4	70.1

GROUP 1	ASTM A335-P1	.50% Moly	GROUP 4	Stainless Steels Types 304, 309, 310, 316 (including "L" & "H" Grades)
	ASTM A335-P2	.50-.70% Chr.		
	ASTM A335-P12	.85-1.10% Chr.		
	ASTM A335-P11	1.05-1.45% Chr.	GROUP 5	Copper, Brass, Everdur
	ASTM A335-P3	1.50-2.00% Chr.		
GROUP 2	ASTM A335-P3b	1.75-2.25% Chr.	GROUP 6	ASTM A-333-GR-1 ASTM A-333-GR4 ASTM A-333-GR9
	ASTM A335-P22	2.00-2.50% Chr.		
	ASTM A335-P21	2.75-3.25% Chr.		
	ASTM A335-P5,b,c	4.00-6.00% Chr.		
GROUP 3	ASTM A335-P7	6.00-8.00% Chr.	GROUP 7	Hastelloy, Titanium, 99% Ni.
	ASTM A335-P9	8.00-10.00% Chr.	GROUP 8	Stainless Steels Types 321 & 347, Cu-Ni, Monel, Inconel, Incoloy, Alloy 20
	Ferritic Chromes	10.00-15.00% Chr.		
	ASTM A333 Gr. 3	3.50% Nickel	GROUP 9	Aluminum

Fabrication of alloy and non-ferrous piping is to be figured on the same basis as for corresponding operation on carbon steel materials, plus the percentage given below applicable to the carbon steel man hour schedules.

HYDROSTATIC TESTING (PERCENT ADDITIVE)

Nominal Pipe Size	MATERIAL CLASSIFICATION—GROUP NUMBERS AND PERCENTAGES								
	Group 1	Group 2	Group 3	Group 4	Group 5	Group 6	Group 7	Group 8	Group 9
2 or less	5.0	7.0	9.0	9.5	4.0	11.5	25.0	11.0	14.0
3	5.5	7.0	10.0	10.0	4.5	12.0	26.0	11.5	15.0
4	6.0	8.0	11.0	11.5	5.0	13.5	27.0	12.0	16.5
5	6.5	8.5	11.5	12.0	5.5	14.5	--	12.5	17.0
6	7.0	9.0	12.0	13.0	6.0	15.0	28.0	13.0	19.0
8	8.0	10.0	14.0	14.0	10.0	17.5	30.0	15.0	21.0
10	9.0	11.0	15.0	16.0	15.0	19.0	35.0	17.0	23.0
12	9.5	12.0	16.0	17.0	16.0	21.0	40.0	19.0	25.0
14	10.0	13.0	17.5	19.0	20.0	23.0	--	20.0	27.0
16	10.5	14.0	19.0	20.0	22.0	25.5	--	23.0	29.0
18	12.0	16.0	21.0	22.0	23.0	27.5	--	24.5	32.0
20	13.0	17.0	23.0	25.0	25.0	30.0	--	26.0	36.0
22	14.0	18.5	25.0	26.0	27.0	31.0	--	28.0	38.0
24	15.0	20.0	26.0	28.0	28.0	32.0	--	29.0	40.0
26	16.4	21.6	28.1	30.4	30.4	34.6	—	31.2	43.4
28	17.6	23.2	30.2	32.8	32.8	37.2	—	33.6	46.8
30	18.9	24.9	32.4	35.1	35.1	39.9	—	36.0	50.1
32	20.2	26.6	34.6	37.4	37.4	42.6	—	38.4	53.4
34	21.4	28.2	36.7	39.8	39.8	45.2	—	40.8	56.8
36	22.7	29.9	38.9	42.1	42.1	47.9	—	43.2	60.1
38	23.9	31.5	41.0	44.5	44.5	50.5	—	45.6	63.5
40	25.2	33.2	43.2	46.8	46.8	53.2	—	48.0	66.8
42	26.5	34.9	45.4	49.1	49.1	55.9	—	50.4	70.1

	ASTM A335-P1	.50% Moly	GROUP 4	Stainless Steels	
	ASTM A335-P2	.50-.70% Chr.		Types 304, 309, 310, 316 (including "L" & "H" Grades)	
GROUP 1	ASTM A335-P12	.85-1.10% Chr.			
	ASTM A335-P11	1.05-1.45% Chr.	GROUP 5	Copper, Brass, Everdur	
	ASTM A335-P3	1.50-2.00% Chr.			
	ASTM A335-P3b	1.75-2.25% Chr.	GROUP 6	ASTM A-333-GR-1	
	ASTM A335-P22	2.00-2.50% Chr.		ASTM A-333-GR4	
GROUP 2	ASTM A335-P21	2.75-3.25% Chr.		ASTM A-333-GR9	
	ASTM A335-P5,b,c	4.00-6.00% Chr.			
	ASTM A335-P7	6.00-8.00% Chr.	GROUP 7	Hastelloy, Titanium, 99% Ni.	
	ASTM A335-P9	8.00-10.00% Chr.	GROUP 8	Stainless Steels Types 321 & 347, Cu-Ni, Monel, Inconel, Incoloy, Alloy 20	
GROUP 3	Ferritic Chromes	10.00-15.00% Chr.			
	ASTM A333 Gr. 3	3.50% Nickel	GROUP 9	Aluminum	

Fabrication of alloy and non-ferrous piping is to be figured on the same basis as for corresponding operation on carbon steel materials, plus the percentage given below applicable to the carbon steel man hour schedules.

ACCESS HOLES

Percent Additive

Nominal Pipe Size	MATERIAL CLASSIFICATION—GROUP NUMBERS AND PERCENTAGES								
	Group 1	Group 2	Group 3	Group 4	Group 5	Group 6	Group 7	Group 8	Group 9
2-1/2, 3, 4	6.0	8.0	11.0	11.5	5.0	13.5	27.0	12.0	16.5
5, 6, 8	8.0	10.0	14.0	14.0	10.0	17.5	30.0	15.0	21.0
10, 12	9.5	12.0	16.0	17.0	16.0	21.0	40.0	19.0	25.0
14, 16, 18	12.0	16.0	21.0	22.0	23.0	27.5	--	24.5	32.0
20, 22, 24	15.0	20.0	26.0	28.0	28.0	32.0	--	29.0	40.0

GROUP 1	ASTM A335-P1	.50% Moly	GROUP 4	Stainless Steels Types 304, 309, 310, 316 (including "L" & "H" Grades)
	ASTM A335-P2	.50-.70% Chr.		
	ASTM A335-P12	.85-1.10% Chr.		
	ASTM A335-P11	1.05-1.45% Chr.	GROUP 5	Copper, Brass, Everdur
	ASTM A335-P3	1.50-2.00% Chr.		
GROUP 2	ASTM A335-P3b	1.75-2.25% Chr.	GROUP 6	ASTM A-333-GR-1 ASTM A-333-GR4 ASTM A-333-GR9
	ASTM A335-P22	2.00-2.50% Chr.		
	ASTM A335-P21	2.75-3.25% Chr.		
	ASTM A335-P5,b,c	4.00-6.00% Chr.		
GROUP 3	ASTM A335-P7	6.00-8.00% Chr.	GROUP 7	Hastelloy, Titanium, 99% Ni.
	ASTM A335-P9	8.00-10.00% Chr.	GROUP 8	Stainless Steels Types 321 & 347, Cu-Ni, Monel, Inconel, Incoloy, Alloy 20
	Ferritic Chromes	10.00-15.00% Chr.		
	ASTM A333 Gr. 3	3.50% Nickel	GROUP 9	Aluminum

Section Four

PNEUMATIC MECHANICAL INSTRUMENTATION

This section is included to cover the complete man hours required for installing pneumatic mechanical instrumentation, as may be required for the monitoring of various process systems.

The man hours listed are for labor only and do not have any bearing on material or equipment cost.

All labor man hours are included for unloading, from railroad cars or trucks, hauling to and unloading at job storage facilities, hauling from storage to erection site, calibrating when necessary, positioning in place, testing, and final check.

LIQUID LEVEL GAUGE GLASSES

Transparent Type

MAN HOURS EACH

Visible Length Inches	Man Hours Each	
	CS	316SS
10-1/4	5.72	6.29
12-5/8	6.93	7.62
19-3/4	7.59	8.34
26-3/4	7.70	8.47
33-3/4	8.25	9.07
45-1/2	8.91	9.80
55	9.46	10.40
65-3/8	10.78	11.85
78-3/4	11.99	13.18

Note: CS = carbon steel, SS = stainless steel

Above rating 2,000 psi @ 100° F to 375 psi @ 600° F.

Man hours are based on Penberthy TL Series.

Man hours include checking out of storage, calibrating if necessary, hauling to location and installing. Man hours do not include valve, piping, or any electrical installation. If liquid level gauge valves are required, use 2.0 man hours per pair.

LIQUID LEVEL GAUGE GLASSES

Transparent Type

MAN HOURS EACH

Visible Length Inches	Man Hours Each	
	CS	316SS
6-3/4	5.06	5.56
10-1/4	5.72	6.29
12-5/8	6.93	7.62
19-3/4	7.59	8.34
26-3/4	8.25	9.07
33-3/4	8.91	9.80
45-1/2	10.12	11.13
55	11.99	13.18
65-3/8	12.65	13.91
78-3/4	13.97	15.36

Note: CS = carbon steel, SS = stainless steel

Above rating 800 psi @ 100° F to 450 psi @ 600° F.

Man hours are based on Penberthy TLC Series.

Man hours include checking out of storage, calibrating if necessary, hauling to location and installing. Man hours do not include valve, piping, or any electrical installation. If liquid level gauge valves are required, use 2.0 man hours per pair.

LIQUID LEVEL GAUGE GLASSES

Transparent Type

MAN HOURS EACH

Visible Length Inches	Man Hours Each	
	CS	316SS
6-3/4	5.06	5.56
10-1/4	5.72	6.29
12-5/8	6.93	7.62
19-3/4	7.59	8.34
26-3/4	8.25	9.07
33-3/4	8.91	9.80
45-1/2	10.12	11.13
55	11.99	13.18
65-3/8	12.65	13.91
78-3/4	13.97	15.36

Note: CS = carbon steel, SS = stainless steel

Above rating 2,500 psi @ 100° F to 750 psi @ 600° F.

Man hours are based on Penberthy TM Series.

Man hours include checking out of storage, calibrating if necessary, hauling to location and installing. Man hours do not include valve, piping, or any electrical installation. If liquid level gauge valves are required, use 2.0 man hours per pair.

LIQUID LEVEL GAUGE GLASSES

Transparent Type

MAN HOURS EACH

Visible Length Inches	Man Hours Each	
	CS	316SS
6-3/4	4.40	4.84
10-1/4	5.06	5.56
12-5/8	6.38	7.01
19-3/4	6.93	7.62
26-3/4	6.98	7.67
33-1/4	7.59	8.34
45-1/2	8.25	9.07
55	8.91	9.80
65-3/8	10.12	11.13
78-3/4	11.44	12.58

Note: CS = carbon steel, SS = stainless steel

Above rating 3,000 psi @ 100° F to 1,500 psi @ 600° F.

Man hours are based on Penberthy TH Series.

Man hours include checking out of storage, calibrating if necessary, hauling to location and installing. Man hours do not include valve, piping, or any electrical installation. If liquid level gauge valves are required, use 2.0 man hours per pair.

LIQUID LEVEL GAUGE GLASSES

Reflex Type

MAN HOURS EACH

Visible Length Inches	Man Hours Each	
	CS	316SS
10-1/4	5.06	5.56
12-5/8	6.38	7.01
19-3/4	6.93	7.62
26-3/4	6.98	7.67
33-3/4	7.59	8.34
45-1/2	8.25	9.07
55	8.91	9.80
65-3/8	10.12	11.13
78-3/4	11.44	12.58

Note: CS = carbon steel, SS = stainless steel

Above rating 2,400 psi @ 100° F to 1,300 psi @ 600° F.

Man hours are based on Penberthy RL Series.

Man hours include checking out of storage, calibrating if necessary, hauling to location and installing. Man hours do not include valve, piping, or any electrical installation. If liquid level gauge valves are required, use 2.0 man hours per pair.

LIQUID LEVEL GAUGE GLASSES

Reflex Type

MAN HOURS EACH

Visible Length Inches	Man Hours Each	
	CS	316SS
6-3/4	4.40	4.84
10-1/4	5.06	5.56
12-5/8	6.38	7.01
19-3/4	6.93	7.62
26-3/4	6.98	7.67
33-3/4	7.59	8.34
45-1/2	8.25	9.07
55	8.91	9.80
65-3/8	10.12	11.13
78-3/4	11.44	12.58

Note: CS = carbon steel, SS = stainless steel

Above rating 2,400 psi @ 100° F to 550 psi @ 600° F.

Man hours are based on Penberthy RLC Series.

Man hours include checking out of storage, calibrating if necessary, hauling to location and installing. Man hours do not include valve, piping, or any electrical installation. If liquid level gauge valves are required, use 2.0 man hours per pair.

LIQUID LEVEL GAUGE GLASSES

Reflex Type

MAN HOURS EACH

Visible Length Inches	Man Hours Each	
	CS	316SS
6-3/4	4.43	4.85
10-1/4	5.10	5.60
12-5/8	6.40	7.04
19-3/4	6.95	7.65
26-3/4	7.00	8.00
33-3/4	7.63	8.37
45-1/4	8.30	9.10
55	8.91	9.85
65-3/8	10.12	11.17
78-3/4	11.44	12.62

Note: CS = carbon steel, SS = stainless steel

Above rating 3,000 psi @ 100° F to 1,700 psi @ 600° F.

Man hours are based on Penberthy RM Series.

Man hours include checking out of storage, calibrating if necessary, hauling to location and installing. Man hours do not include valve, piping, or any electrical installation. If liquid level gauge valves are required, use 2.0 man hours per pair.

LIQUID LEVEL GAUGE GLASSES

Reflex Type

MAN HOURS EACH

Visible Length Inches	Man Hours Each	
	CS	316SS
6-3/4	4.43	4.85
10-1/4	5.10	5.60
12-5/8	6.40	7.04
19-3/4	6.95	7.65
26-3/4	7.00	8.00
33-3/4	7.63	8.37
45-1/2	8.30	9.10
55	8.91	9.85
65-3/8	10.12	11.17
78-3/4	11.44	12.62

Note: CS = carbon steel, SS = stainless steel
Above rating 4,000 psi @ 100° F to 2,200 psi @ 600° F.
Man hours are based on Penberthy RH Series.
Man hours include checking out of storage, calibrating if necessary, hauling to location and installing. Man hours do not include valve, piping, or any electrical installation. If liquid level gauge valves are required, use 2.0 man hours per pair.

Pressure Gauges

MAN HOURS EACH

Pressure Dial Size Inches	Man Hours Each For			
	Type 1	Type 2	Type 3	Type 4
2½	0.55	—	—	—
3½	—	0.55	0.84	1.20
4½	—	0.66	0.92	1.32
8½	—	0.92	1.19	1.45

Type 1: Brass bourdon tube, drawn steel case, black enamel finish, ¼″ bottom connection, 0–30 through 0–600 psi range. Accuracy: middle half of scale 2% of scale range; remainder 3%.

Type 2: Brass bourdon tube, phenol turret case, black finish, ¼″ bottom connection, 0–15 through 0–1,000 psi range. Accuracy 1% of scale.

Type 3: Drawn stainless steel bourdon tube, stainless movement, phenol turret case, ¼″ bottom connection, 0-15 through 0–1,000 psi range. Accuracy ½ of 1%.

Type 4: Alloy steel bourdon tube, bronze rotary movement, cast aluminum case, ¼″ bottom connection, 0–1,000 through 0–10,000 psi range. Accuracy 1%.

Man hours include checking out of storage, calibrating, hauling to erection site, installing, testing, and final check.

Man hours do not include installation of piping. See piping section for these charges.

PNEUMATIC LIQUID LEVEL INSTRUMENTS

Local Mounted

MAN HOURS EACH

Displacer Length Inches	Man Hours Each		
	Type 1	Type 2	Type 3
14	3.9	3.9	3.9
32	5.1	5.1	5.1
48	6.4	6.4	6.4
60	7.6	7.6	7.6
72	8.9	8.9	8.9
84	10.1	10.1	10.1
96	11.4	11.4	11.4
108	12.3	12.3	12.3
120	14.0	14.0	14.0

Note: LT or LC: Level transmitter or controller, pneumatic, side mounted, external displacement type. Fabricated steel cage.

Fisher Type 2500 pneumatic controller direct or reverse acting; proportional band adjustment; two 2″ φ 30 psi gauges with airset mounted on a fabricated steel displacer cage. 316 SS trim; 304 SS displacer; K-Monel torque tube.

Type 1: 600# screwed connection; 1½″ or 2″; fabricated steel cage; top and bottom connections; Fisher Type 249A.
Type 2: 600# flanged connections; 1½″ or 2″; fabricated steel cage; top and bottom connections; Fisher Type 249A.
Type 3: 300# flanged connections; 1½″ or 2″; fabricated steel cage; top and bottom connections; Fisher Type 249A.

Above man hours include checking out of storage, calibrating, hauling to erection site, installing, testing, and final checking of the level transmitters or controllers complete with air supply filter-regulator supply and output gauges.

Man hours do not include installation of air supply and air signal lines or bolt-up or make-on of flanges. See other piping accounts for these man hours.

PNEUMATIC LIQUID LEVEL INSTRUMENTS

Local Mounted

MAN HOURS EACH

Displacer Length Inches	Man Hours Each				
	Type 1	Type 2	Type 3	Type 4	Type 5
14	5.1	5.1	5.2	5.2	5.7
32	5.7	5.7	5.9	5.9	6.4
48	6.4	6.4	6.6	6.6	-
60 thru 96	8.2	8.2	8.6	8.6	-
108	8.3	8.3	8.7	8.7	-
120	9.0	9.0	9.2	9.2	-

Note: LT or LC: Level transmitter or controller, pneumatic, top mounted, internal displacer.

Fisher Type 2500 pneumatic controller direct or reverse acting; proportional band adjustment; two 2" φ 30 psi gauges with airset mounted on a top mounted displacer assembly. 316 SS trim; 304 SS displacer; K-Monel torque tube.

Type 1: 4" 125# flanged; cast iron head; Fisher Type 249A.
Type 2: 4" 250# flanged; cast iron head; Fisher Type 249A.
Type 3: 8" 125# flanged; cast iron head; Fisher Type 249A.
Type 4: 8" 250# flanged; cast iron head; Fisher Type 249A.
Type 5: 4" 900# flanged; cast steel head; Fisher Type 249P.

Above man hours include checking out of storage, calibrating, hauling to erection site, installing, testing, and final checking of the level transmitters or controllers complete with air supply filter-regulator supply and output gauges.

Man hours do not include installation of air supply and air signal lines or bolt-up of flanges. See other piping accounts for these man hours.

PNEUMATIC PRESSURE INSTRUMENTS

Local Mounted

MAN HOURS EACH

Item	Man Hours Each
PT gauge pressure transmitter, pneumatic. Foxboro Model: 11GM. Range: spans from 10–2,000 psi with maximum range of 3,000 psi within limits of range capsule, maximum overrange 4,000 psi. Materials: 316 SS. Process Connection: ¼″ or ½″ NPT female. Output Signal: 3–15 psi. Mounting: Bracket for 2″ pipe. With air filter-regulator set and mounting bracket.	7.6
PT absolute pressure transmitter, pneumatic. Foxboro Model: 11AH. Range Capsule: 20–200 psi, adjustable span, maximum overrange 350 psi. Body Material: 316 SS. Process Connection: ½″ NPT female. Output Signal: 3–15 psi. Mounting: Bracket for 2″ pipe. With air filter-regulator set and mounting bracket.	7.6
PTI pressure indicating transmitter, pneumatic indicating. FoxboroModel: 45P. Case: Rectangular. Range: 0–10 to 0–2,000 psi. Scale: Eccentric, 6-⅛″ length. Output Signal: 3–15 psi. Pressure Element: Cu-Ni-Mn, Diaphragm. With air filter-regulator set and mounting yoke.	5.1
PR pressure recorder direct connected. Foxboro Model: 40PR. Case: Rectangular. Mounting: Yoke, Chart Drive: Electric, 115 volts, 60 Hz., 24-hour. Range: 0–10 to 0–2,000 psi. Pen: One. Pressure Element: Cu-Ni-Mn, diaphragm. With mounting yoke.	15.2
PC pressure controller, pneumatic, direct connected. Foxboro Model: 43AP-FA4. Control Function: Proportional plus reset. Prop. Band: 4–400%. Reset Time: 0.5–25 minutes. Range: 0–10 to 0–2,000 psi. Element: Cu-Ni-Mn, Diaphragm. Scale: Eccentric. Relay Action: Reversible. Set Point Knob: Internal. Output Gauge: 0–30 psi. Mounting: Bracket for 2″ pipe. With air filter-regulator set and mounting bracket.	15.2

Above man hours include checking out of storage, calibrating, hauling to erection site, installing, testing, and final checkout.

Man hours do not include connections to process and air signal lines. See piping accounts for these charges.

Man hours do not include wiring for recorder electric charge drive. See *Electrical Man Hour Manual* for this charge.

PNEUMATIC TEMPERATURE INSTRUMENTS

Local Mounted

MAN HOURS EACH

Item	Man Hours Each
TT temperature transmitter, pneumatic, Foxboro Model: 12A. Thermal System: Gas pressure, Class IIIB. Range: -100 to +1,000° F. Bulb: 316 SS, adjustable union with bendable 18" extension ⅜" diameter, insertion adjustable, 8–21". Bushing: ¾" NPT. Tubing: 3½" vinyl-covered flexible SS protection over SS capillary. Output Signal: 3–15 psi. Mounting: Universal bracket for surface or 2" pipe. With air filter-regulator set and mounting bracket.	7.6
TT temperature transmitter, pneumatic non-indicating, Foxboro Model: 44BT. Mounting: Bracket for surface or 2" pipe. Connections: Bottom. Range: -100 to +1,000° F. Thermal System: Class IIIB, gas pressure. Bulb: 316 SS, plain bulb. Tubing: 5'⅛" OD, SS. Output Signal: 3–15 psi. With air filter-regulator set and mounting bracket.	7.6
TTI temperature transmitter, pneumatic indicating, Foxboro Model: 45P. Case: Rectangular. Range: -300 to +600° F. Scale: Eccentric 6-1/8" length. Output Signal: 3–15 psi. Thermal System: Class IIIB, gas pressure. Bulb: 316 SS, fixed union with 8" bendable extension. Tubing: 5'-1/8" OD, SS. With air filter-regulator set and mounting yoke.	7.6
TR temperature recorder, direct connected. Foxboro Model: 40 PR. Case: Rectangular. Mounting: Yoke. Connection: Bottom. Pen: One, V-type. Chart Drive: Electric, 115 volts, 60 Hz, 24-hour. Thermal System: Class III, Gas Pressure. Bulb: 316 SS, fixed union with 8" bendable extension. Bushing: Plain, 316 SS, ¾ NPT. Tubing: 5'⅛" OD, 316 SS, with mounting yoke.	15.2
TC temperature controller, pneumatic, direct connected. Foxboro Model: 43AP-FA4. Control Function: Proportional plus reset. Prop. Band: 4–400%. Reset Time: 0.5–25 minutes. Thermal System: Class III, gas pressure. Bulb: 316 SS, fixed union with 8" bendable extension. Tubing: 5'⅛" OD 316 SS. Bushing: Plain, 316 SS, ¾" NPT. With air filter-regulator set and mounting yoke.	17.7

Above man hours include checking out of storage, calibrating, hauling to erection site, installing, including connecting capillary bulb to process, testing, and final check.

Man hours do not include installing air supply or air signal lines. See other piping accounts for these time frames.

Man hours do not include electrical installation for electrical hook-up required for chart driver on the recorders. See *Electrical Man Hour Manual* for these time frames.

THERMOMETERS AND THERMOWELLS

MAN HOURS EACH

Thermometer Description	Man Hour Each
Straight form 7″ scale, 3½″ stem	0.77
Straight form 9″ scale, 6″ stem	0.84
Straight form 12″ scale, 12″ stem	0.92
Angle form 7″ scale, 3½″ stem	0.77
Angle form 9″ scale, 6″ stem	0.84
Angle form 12″ scale, 12″ stem	0.92
Every angle adjustable, 7″ scale, 3½″ stem	0.77
Every angle adjustable, 9″ scale, 6″ stem	0.84
Every angle adjustable, 12″ scale, 12″ stem	0.92

Man hours are for thermometers, mercury in glass type, epoxy-coated aluminum case, red-reading mercury; -40° F minimum to 750° F maximum range (600° F maximum temperature differential span), stainless steel stem and including separable 316 stainless steel well with ¾″ NPT (external thread).

Man hours include checking out of storage, hauling to erection site, installing, checking, and testing.

Man hours do not include pipe or pipe fittings. See piping tables for these man hours.

THERMOMETERS AND THERMOWELLS

MAN HOURS EACH

Thermometer Description	Man Hour Each
Bottom Connection, 3″ Dial, 4″ Stem	0.77
Bottom Connection, 5″ Dial, 6″ Stem	0.84
Bottom Connection, 5″ Dial, 12″ Stem	0.84
Back Connection, 3″ Dial, 4″ Stem	0.77
Back Connection, 3″ Dial, 6″ Stem	0.77
Back Connection, 5″ Dial, 12″ Stem	0.84
Every Angle Adjustable, 5″ Dial, 4″ Stem	0.84
Every Angle Adjustable, 5″ Dial, 6″ Stem	0.84
Every Angle Adjustable, 5″ Dial, 12″ Stem	0.84

Man hours are for thermometers, dial type, bi-metal, 18-8 stainless steel case and stem, scale length approximately two times dial diameter; - 80° F minimum to 500° F maximum (500° F maximum temperature differential span); and including separable 316 stainless steel well with ¾″ NPT (external thread).

Man hours include checking out of storage, hauling to erection site, installing, checking, and testing.

Man hours do not include pipe or pipe fittings. See piping tables for these man hours.

THERMOMETERS AND THERMOWELLS

MAN HOURS EACH

Thermometer Description	Man Hours Each
4½″ Dial, 3½″ Immersion on Bulb	1.10
6″ Dial, 6″ Immersion on Bulb	1.20
8″ Dial, 12″ Immersion on Bulb	1.30

Above man hours are for thermometers, dial type, remote reading, filled system type, mercury actuated, lower or back connected phenolic case for wall or flush mounting, stainless steel movement, complete with 18-8 stainless steel bulb, flexible extension and 5 lineal feet of copper capillary; -20° F minimum to 1,000° F maximum range, (800° F maximum temperature differential span). ¾″ NIP union connection including 316 stainless steel well with ¾″ NPT (external thread).

Man hours include checking out of storage, hauling to erection site, installing, checking, and testing.

Man hours do not include pipe or pipe fittings. See piping tables for these man hours.

THERMOWELLS AND THERMOCOUPLES

MAN HOURS EACH

Thermowells and Thermocouples	Man Hours Each
Thermowells: ¾″ × ½″ NPT 304 SS	
Thermowell Length— 3½″	0.7
6″	0.7
8″	0.7
10″	0.7
12″	0.7
18″	0.7
24″	0.7
Thermowells: 1″ × ½″ NPT 304 SS	
Thermowell Length— 3½″	0.9
6″	0.9
8″	0.9
10″	0.9
12″	0.9
18″	0.9
24″	0.9
Thermowells: 150# Flanged 304 SS	
Thermowell Length— 6″	1.0
8″	1.0
10″	1.0
12″	1.0
18″	1.0
Thermocouple head ½″ NPT 304 SS 6″ long	0.7

Note: SS = stainless steel

For 300# flanged add 15% to above man hours

Man hours include checking out of storage, hauling to erection site, and installing.

RELIEF VALVES

MAN HOURS EACH

Type	Size Inches	Man Hour Each
A	½ × 1	0.9
A	¾ × 1	1.0
A	1 × 1	1.0
B	½ × 1	0.9
B	¾ × 1	1.0
B	1 × 1	1.0

Type A: Screwed relief valve, carbon steel body, carbon steel spring, pressure limit 2,000#, maximum temperature 600° F.

Type B: Screwed relief valve, carbon steel body, alloy steel spring, pressure limit 2,000#, maximum temperature 750° F.

Man hours include checking out of storage, hauling to erection site, installing, testing, and checking.

Man hours do not include piping to or from the valve or supports if required. See other piping tables for these requirements.

FLANGED RELIEF VALVES

MAN HOURS EACH

Inlet and Outlet Size Inches	Man Hours Each Flange Rating		
	150#	300#	600#
1 × 2	1.4	1.6	1.8
1½ × 2	1.5	1.6	1.8
1½ × 2½	1.5	1.7	1.9
1½ × 3	1.5	—	—
2 × 3	1.6	1.7	1.9
2½ × 4	—	2.3	2.5
3 × 4	2.0	2.3	2.5
4 × 6	2.7	3.1	3.3
6 × 8	3.6	4.1	4.4
6 × 10	—	4.7	5.0
8 × 10	4.8	5.4	5.8

Above are standard relief valves A-216 grade WCB carbon steel body, closed bonnet with screwed cap, carbon steel spring. Maximum temperature 450° F.

Man hours include checking out of storage, hauling to erection site, installing, testing, and checking out.

Man hours do not include piping to or from the valve or supports if required. See other piping tables for these requirements.

FLANGED RELIEF VALVES

MAN HOURS EACH

Inlet and Outlet Size Inches	Man Hour Each Flange Rating	
	300#	600#
1 × 2	1.6	1.8
1½ × 2	1.6	1.8
1½ × 2½	1.7	1.9
2 × 3	1.7	1.9
2½ × 4	2.3	2.5
3 × 4	2.3	2.5
4 × 6	3.1	3.3
6 × 8	4.1	—
6 × 10	4.7	5.0
8 × 10	5.4	—

Above are standard relief valves, A-216 grade WCB carbon steel body, closed bonnet with screwed cap, tungsten steel spring. Maximum temperature 800° F.

Man hours include checking out of storage, hauling to erection site, installing, testing, and checking.

Man hours do not include piping to or from the valve or supports if required. See other piping tables for these requirements.

PNEUMATIC FLOW TRANSMITTERS

MAN HOURS EACH

Item	Man Hours Each
FT flow transmitter, pneumatic, D/P cell. Foxboro Model: 15-A1— for static pressure to 500 psig. Range Capsule: 5–25″ water, adjustable span. Body Material: Cadmium-plated carbon steel. Process Connection: ¼″ NPT or ½″ NPT female of ½″ Sch. 80 welding neck. Output Signal: 3–15 psi. Mounting: Direct to process or by bracket for 2″ pipe. With air filter-regulator set and mounting bracket.	7.6
FT flow transmitter, pneumatic, D/P cell. Foxboro Model: 13-A1— for static pressure to 1,500 psig. Range Capsule: 20–205″ water or 200–850″ water, adjustable span. Body Material: Cadmium-plated carbon steel. Process Connection: ¼″ NPT or ½″ NPT female of ½″ Sch. 80 welding neck. Output Signal: 3–15 psi. Mounting: Direct to process or by bracket for 2″ pipe. With air filter-regulator set and mounting bracket.	7.6
FT flow transmitter, pneumatic, D/P cell. Foxboro Model: 13H-A1— for static pressure to 6,000 psig. Range Capsule: 20–205″ water or 200–850″ water, adjustable span. Body Material: Cadmium-plated carbon steel. Process Connection: ¼″ NPT or ½″ NPT female. Output Signal: 3–15 psi. Mounting: Direct to process or by bracket for 2″ pipe. With air filter-regulator set and mounting bracket.	7.6

Above man hours include checking out of storage, calibrating, hauling to erection site, installing, testing, and final check.

Man hours do not include connections to process supply and air signal lines. See piping accounts for these man hours.

FLOW INDICATING TRANSMITTERS, FLOW RECORDERS, AND FLOW CONTROLLERS

MAN HOURS EACH

Item	Man Hours Each
FTI Flow Indicating Transmitter (mercury-less type), D/P cell type. Foxboro Model: 45P. Case: Rectangular. Mounting: Yoke. Scale: Eccentric, 6-⅛″ length. Output Signal: 3–15 psi. Meter Body & Covers: Type 37, forged steel, cadmium-plated. Diaphragm: Stainless steel. Differential Range: 0–50″ to 0–200″ of water. Connections: ¼″ or ½″; top or bottom. With air filter-regulator set and mounting yoke.	12.7
FC Flow Indicating Controller, Direct Connected, D/P Cell Type. Foxboro Model: 43AP-FA 4. Control Function: Proportional plus reset. Prop. Band: 4–400%. Reset Time: 0.5–25 minutes. Meter Body & Covers: Type 37 forged steel, cadmium-plated. Differential Range: 0–50″ to 0–200″ of water. Scale: Eccentric. Ready Action: Reversible. Set Point Knob: Internal. Output Gauge: 0–30 psi. Mounting: Yoke. Connections: ¼″ or ½″; top or bottom. With air filter-regulator set and mounting yoke.	17.7
FR Flow Recorder, Direct Connected, D/P Cell Type. Foxboro Model: 40PR. Case: Rectangular. Mounting: Yoke. Pen: One, V-type. Chart Drive: Electric, 115 volts, 60 Hz, 24-hour. Meter Body & Covers: Type 37, forged steel, cadmium-plated. Differential Range: 0–50″ to 0–200″ of water. Connections: ¼″ or ½″; top or bottom with mounting yoke.	15.2

Above man hours include checking out of storage, calibrating, hauling to erection site, installing, testing and final check.

Man hours do not include connections to process air supply and air signal lines. See piping account for mechanical man hours. See *Electrical Man Hour Manual* for electrical charges.

PNEUMATIC LIQUID LEVEL TRANSMITTERS

Local Mounted

MAN HOURS EACH

Item	Man Hours Each
LT Liquid Level Transmitter, Pneumatic, D/P Cell. Foxboro Model: 15FA1. Range Capsule: 5–25″ water, adjustable span. Body & Flange: Cadmium-plated carbon steel. Process Connection: High pressure-ANSI raised face modified flange; Low pressure - ½ NPT. Flange: 6″; 150-lb with 5″ flange extension. Output Signal: 3–15 psi. Mounting: By flange process connection. With air filter-regulator set.	7.6
LT Liquid Level Transmitter, Pneumatic, D/P Cell. Foxboro Model: 13FA1. Range Capsule: 20–205″ water or 200–850″ water, adjustable span. Body & Flange: Cadmium-plated carbon steel. Process Connection: High pressure-ANSI raised face modified flange; Low pressure - ½ NPT. Flange: 3″; 150-lb with 5″ flange extension. Output Signal: 3–15 psi. Mounting: By flange process connection. With air filter-regulator set.	7.6

Above man hours include checking out of storage, calibrating, hauling to erection site, installing, testing and final checking of the level transmitters or controllers complete with air supply filter-regulator, supply, and output gauges.

Man hours do not include installation of air supply and air signal lines or bolt-up of flanges. See other piping accounts for these man hours.

Control Panel Installation

Control panels are usually fabricated by a sub-contractor who specializes in this type of work. The instruments that are to be installed on the panel board or cabinet are usually furnished to the sub-contractor by the general contractor. Panel boards are usually fabricated in sections up to approximately 12'0" in length.

To unload control panel from carrier, move into position, and set on foundation. Per linear foot of control board length . . . . 1.5 man hours

If more than one section of control panel is required,
add an additional time for each connection . . . . 2.5 man hours

CONNECTING PNEUMATIC PANEL BOARD INSTRUMENTS

MAN HOURS EACH

Description	Man Hours Each
Recording Controller—large case 12″ circular chart, with automatic reset.	
one pen, one controller	4.4
two pens, one controlling, one recording only	6.6
two pens, two controllers	8.8
three pens, two controlling, one recording only	11.0
Receiving Recorder—large case 12″ circular chart, one pen	2.7
two pens	3.3
three pens	5.0
Receiving Indicator—concentric dial 3-½″ circular case	1.7
ribbon type, horizontal or vertical 4″ scale	1.7
ribbon type, for two variables	3.3
Manual Loading Station—6″ × 5-¼″ with circular dial and horizontal output indicator:	
one pointer diaphragm element	2.2
two pointers diaphragm element	3.3
two pointers diaphragm element with set point regulator	4.4
two pointers diaphragm element with set point regulator and auto./man. unit	5.0
Recorder — 7-½″ × 7-½″, 4″ strip chart, one pen	2.2
two pens	3.3
three pens	5.0
Recording Control Station—7-½″ × 7-½″, 4″ strip chart and vertical output gauge:	
one pen	4.4
one pen, one pointer	4.4
two pens	8.8
two pens, one pointer	8.8
three pens	13.2
Controller—Shelf Mounting: proportional control	3.3
proportional control plus derivative control	4.4
proportional control plus reset control	4.4
proportional control plus reset plus derivative control	4.4

Above man hours include making tubing terminal connections, calibrating, checking, adjusting, testing, and commissioning pneumatic instruments.

Man hours for installation of air supply piping and air signal lines from control panel to remote instruments are not included. See other accounts for these man hours.

No manufacturer's representatives are included in the above man hours.

CONNECTING PNEUMATIC PANEL BOARD INSTRUMENTS

MAN HOURS EACH

Description	Man Hours Each
Controller with Local/Remote Set Point—shelf mounting, proportional control	3.9
proportional control plus derivative control	5.0
proportional control plus reset control	5.0
proportional control plus reset plus derivative control	5.0
Manual Loading Stations—shelf-mounted with output indication	2.2
shelf-mounted with input and output indication	2.2
Auto./Manual Switching Station—shelf-mounted with input and output indication	2.2
Ratio Station—shelf-mounted with manual ratio setting	3.3
add for input indication	1.1
Receiver Recorder—shelf-mounted, 2 units wide, roll or scan-fold chart: one pen	2.2
two pens	3.3
three pens	5.0
four pens	6.6
Flow Integrator—6-digit actual flow indicator Multipoint Temperature Recorder, per pt.	2.2
Multipoint Temperature Indicator, per pt.	1.1
Miscellaneous Pneumatic Panel Instruments, Rear of Panel Mounting: Analog computer for multiplying two variables, or dividing one variable by a second one, or extracting square root of one variable, or squaring one variable	3.3
Square Root Extractor, Input 3–15 psi	2.2
Reversing Relay, 1:1 Ratio	1.1
Booster Relay, 1:1 Ratio	1.1
Selector Relay, High or Low of 2 Pressures	1.1
Limit Relay, high limit	1.1
Limit Relay, low limit	1.1

Above man hours include making tubing terminal connections, calibrating, checking, adjusting, testing and commissioning pneumatic instruments.

Man hours for installation of air supply piping and air signal lines from control panel to remote instruments is not included. See other accounts for these man hours.

No manufacturer's representatives are included in the above man hours.

Section Five

UNDERGROUND PIPING

In this section we have tried to cover all labor items related to a complete installation of underground piping.

First of all, the area in which the pipe is to be installed must be excavated. Before an estimate is made on this item it is well to know the kind of soil that may be encountered. For this reason, we have divided soil into five groups according to the difficulty experienced in excavating it. Soils vary greatly in character and no two are exactly alike.

Group 1: LIGHT SOIL — Earth which can be shoveled easily and requires no loosening, such as sand.

Group 2: MEDIUM OR ORDINARY SOILS — Type of earth easily loosened by pick. Preliminary loosening is not required when power excavating equipment such as shovels, dragline scrapers and backhoes are used. This earth is usually classified as ordinary soil and loam.

Group 3: HEAVY OR HARD SOIL — This type of soil can be loosened by pick but this loosening is sometimes very hard to do. It may be excavated by sturdy power shovels without preliminary loosening. Hard and compacted loam containing gravel, small stones and boulders, stiff clay or compacted gravel are good examples of this type.

Group 4: HARD PAN OR SHALE — A soil that has hardened and is very difficult to loosen with picks. Light blasting is often required when excavating with power equipment.

Group 5: ROCK — Requires blasting before removal and transporting. (May be divided into different grades such as hard, soft, or medium.)

For pipe installation we have included man hour tables covering cast iron, concrete and vitrified clay under this section. For carbon steel pipe installation man hours refer to section two of this manual.

In many instances specifications may call for the coating and wrapping of underground pipe. This too, has been covered with a table listing the direct man hours that are required for coating and wrapping various sizes of pipe.

MACHINE EXCAVATION

	NET MAN HOURS PER 100 CUBIC YARDS								
	LIGHT SOIL			MEDIUM SOIL			HEAVY SOIL		
EQUIPMENT	Op. Engr.	Oiler	Labor	Op. Engr.	Oiler	Labor	Op. Engr.	Oiler	Labor
Power Shovel									
1 cubic yard Dipper	1.1	1.1	1.1	2.0	2.0	2.0	2.7	2.7	2.7
3/4 cubic yard Dipper	1.5	1.5	1.5	2.8	2.8	2.8	3.7	3.7	3.7
1/2 cubic yard Dipper	2.0	2.0	2.0	3.7	3.7	3.7	4.9	4.9	4.9
Backhoe									
1 cubic yard Bucket	1.4	1.4	1.4	2.6	2.6	2.6	3.5	3.5	3.5
3/4 Cubic yard Bucket	1.5	1.5	1.5	3.8	3.8	3.8	3.7	3.7	3.7
1/2 cubic yard Bucket	2.0	2.0	2.0	3.7	3.7	3.7	4.9	4.9	4.9
Dragline									
2 cubic yard Bucket	0.7	0.7	0.7	1.3	1.3	1.3	1.7	1.7	1.7
1 cubic yard Bucket	1.1	1.1	1.1	2.0	2.0	2.0	2.7	2.7	2.7
1/2 cubic yard Bucket	2.0	2.0	2.0	3.7	3.7	3.7	4.9	4.9	4.9
Trenching Machine	--	--	--	3.8	--	7.5	4.8	--	9.4

	NET MAN HOURS PER 100 CUBIC YARDS					
	HARD PAN			ROCK		
EQUIPMENT	Op. Engr.	Oiler	Labor	Op. Engr.	Oiler	Labor
Power Shovel						
1 cubic yard Dipper	3.4	3.4	3.4	3.4	3.4	3.4
3/4 cubic yard Dipper	4.6	4.6	4.6	4.6	4.6	4.6
1/2 cubic yard Dipper	6.1	6.1	6.1	6.1	6.1	6.1
Backhoe						
1 cubic yard Bucket	4.4	4.4	4.4	4.4	4.4	4.4
3/4 cubic yard Bucket	4.6	4.6	4.6	4.6	4.6	4.6
1/2 cubic yard Bucket	6.1	6.1	6.1	6.1	6.1	6.1

Man hours are for operational procedures only and do not include equipment rental or depreciation. This must be added in all cases.

Operation includes excavation and dumping on side line or into trucks for hauling but does not include hauling. See pages on hauling for this charge.

For excavations greater than 5'6" add 25% to above man hours.

HAND EXCAVATION

NET LABORER MAN HOURS PER CUBIC YARD

Soil	Excavation	First Lift	Second Lift	Third Lift
Light	General Dry	1.07	1.42	1.89
	General Wet	1.60	2.13	2.83
	Special Dry	1.34	1.78	2.37
Medium	General Dry	1.60	2.13	2.83
	General Wet	2.14	2.85	3.79
	Special Dry	1.07	2.49	3.31
Hard or Heavy	General Dry	2.67	3.55	4.72
	General Wet	3.21	4.27	5.68
	Special Dry	2.94	3.91	5.20
Hard Pan	General Dry	3.74	4.97	6.61
	General Wet	4.28	5.69	7.57
	Special Dry	4.01	5.33	7.09

Man hours include picking and loosening where necessary and placing on bank out of way of excavation or loading into trucks or wagons for hauling away. Man hours do not include hauling or unloading.

ROCK EXCAVATION

Net Man Hours for Drilling, Blasting and Loading
per Cubic Yard of Rock in Place in Ground

Operation	Labor Hours per Cubic Yard		
	Soft	Medium	Hard
Hand Drill, Plug and Feathers	15.0	21.0	30.0
Hand Drill, Blasting	13.0	16.0	22.0
Machine Drill, Plug and Feathers	8.0	11.0	14.0
Machine Drill, Blasting	4.0	6.0	7.0

Man hours are for above described operations only.

For hauling see respective man hour page.

Equipment and materials must be added in all cases.

SHORING AND BRACING TRENCHES

Net Man Hours per 100 Square Feet

Operation	Laborers	Carpenters	Truck Drivers
Placing	3.0	3.0	0.4
Removing	2.5	---	0.4

Man hours include hauling, erecting and stripping.

DISPOSAL OF EXCAVATED MATERIAL

NET MAN HOURS PER 100 CUBIC YARDS

Length of Haul and Truck Capacity	Average Speed 10 M. P. H.		Average Speed 15 M. P. H.		Average Speed 20 M. P. H.	
	Truck Driver	Laborer	Truck Driver	Laborer	Truck Driver	Laborer
3 cubic yard Truck:						
One Mile Haul	15.0	2.8	11.6	2.8	10.5	2.8
Two Mile Haul	21.8	2.8	16.2	2.8	14.0	2.8
Three Mile Haul	28.2	3.0	20.6	3.0	17.3	3.0
Four Mile Haul	36.0	3.0	26.8	3.0	21.0	3.0
Five Mile Haul	41.7	2.5	--	--	--	--
4 cubic yard Truck:						
One Mile Haul	11.3	2.1	8.8	2.3	7.9	2.1
Two Mile Haul	16.2	2.0	12.0	2.0	10.4	2.2
Three Mile Haul	21.6	2.1	15.8	2.3	13.2	2.1
Four Mile Haul	26.4	2.0	18.7	2.0	15.6	2.0
Five Mile Haul	31.3	1.3	--	--	--	--
5 cubic yard Truck:						
One Mile Haul	9.0	1.7	7.0	1.7	6.3	1.7
Two Mile Haul	13.0	1.7	9.7	1.7	8.3	1.7
Three Mile Haul	17.1	1.8	12.3	1.8	10.4	1.7
Four Mile Haul	21.0	2.0	15.0	2.0	12.4	1.6
Five Mile Haul	25.0	1.7	--	--	--	--
8 cubic yard Truck:						
One Mile Haul	5.6	1.0	4.8	1.0	4.0	1.0
Two Mile Haul	8.2	1.0	6.0	1.0	5.2	1.0
Three Mile Haul	10.5	0.9	7.8	0.9	6.5	1.0
Four Mile Haul	13.2	1.0	9.2	1.0	7.6	1.0
Five Mile Haul	15.6	1.0	--	--	--	--

Man hours include round trip for truck, spotting at both ends, unloading and labor for minor repairs and maintenance to vehicle. For loading and excavating see respective man hour listings.

Man hours do not include equipment rental or depreciation. This must be added in all cases.

BACKFILLING AND TAMPING
NET MAN HOURS PER CUBIC YARD

Soil	Hand Shovel Placed	Bulldoze Placed	Tamped 6" Layers
Light	0.60	0.04	--
Medium	0.80	0.07	1.00
Heavy	1.00	0.10	1.20

Man hours for hand shoveling and tamping is that of common labor. Man hours for bulldozer placing is that of operating engineer.

All man hours are based on backfill materials being located within shoveling distance of excavated area.

UNDERGROUND 150 LBS. B. & S. CAST IRON PIPE
LABOR IN MAN HOURS

MAN HOURS PER FOOT		PER MAKE-ON		
		150 Lb. B & S Fittings		
Size Inches	Pipe Set & Align	Lead & Mech. Joint	Cement Joint	Sulphur Joint
4	0.09	0.50	0.35	0.25
6	0.11	0.57	0.37	0.29
8	0.14	0.70	0.50	0.35
10	0.17	0.85	0.60	0.43
12	0.24	1.23	0.95	0.62
14	0.35	1.78	1.25	0.89
16	0.45	2.28	1.60	1.14
18	0.53	2.68	1.89	1.34
20	0.63	3.19	2.24	1.59
24	0.79	4.00	2.81	2.00

Pipe man hours includes handle, haul, set and align in trench.

Fitting man hours includes one make-on.

Man hours must be added for excavation. See respective pages for this charge.

UNDERGROUND VITRIFIED CLAY AND CONCRETE PIPE

LABOR IN MAN HOURS

Size Inches	CONCRETE PIPE (Not Reinforced)		VITRIFIED CLAY PIPE	
	Set & Align Pipe Per Foot	Cement Poured Joint Each	Set & Align Pipe Per Foot	Poured Joint Each
4	0.07	0.20	0.07	0.25
6	0.08	0.25	0.07	0.29
8	0.10	0.32	0.07	0.35
10	0.11	0.39	0.08	0.43
12	0.15	0.50	0.10	0.62
15	0.19	0.75	0.11	0.89
18	0.28	0.95	0.14	1.14
21	0.29	1.15	0.19	1.38
24	0.32	1.25	0.25	1.63
30	0.40	1.56	0.31	2.04
36	0.48	1.88	0.37	2.44
42	0.56	2.19	0.44	2.85
48	0.64	2.50	0.50	3.26
60	0.80	3.13	0.62	4.07

Man hours includes handle, haul, set in trench and align. Man hours for joint or connection of fittings is for one make-up only.

No labor for excavation or backfill is included. Add from respective pages for these charges.

For reinforced concrete pipe add 5% to man hours listed for concrete pipe.

SOCKET CLAMPS FOR CAST IRON PIPE

NET LABOR IN MAN HOURS

Pipe Size Inches	Friction Clamps Complete	Positive Clamps Complete
4	0.25	0.30
6	0.28	0.33
8	0.33	0.38
10	0.38	0.43
12	0.45	0.52
14	0.52	0.62
16	0.60	0.75
18	0.68	0.85
20	0.75	0.95
24	0.88	1.10

Man hours are for labor only and include handling, hauling and the complete installation in all cases.

PIPE COATED WITH TAR AND FIELD WRAPPED BY MACHINE

NET MAN HOURS PER LINEAL FOOT

Nominal Pipe Size	Man Hours Per Foot	Nominal Pipe Size	Man Hours Per Foot
3/4	0.04	22	0.50
1	0.04	24	0.54
1-1/4	0.05	26	0.59
1-1/2	0.06	28	0.63
2	0.07	30	0.68
2-1/2	0.08	32	0.73
3	0.09	34	0.78
4	0.12	36	0.82
5	0.13	38	0.87
6	0.16	40	0.91
8	0.20	42	0.96
10	0.25	44	1.00
12	0.28	46	1.05
14	0.32	48	1.10
16	0.37	54	1.24
18	0.41	60	1.38
20	0.45	--	--

Man hours include:

Sandblast commercially	Apply two ply of 15# tarred felt
Apply one prime coat of pipeline primer	Apply one seal coat of pipeline enamel
Apply 3/32" pipeline enamel	

For hand coating and wrapping add 100% to above man hours.

Section Six

HANGERS
AND
SUPPORTS

The following table is intended to cover labor in man hours for the hanging and/or supporting of a process piping system.

It includes labor man hours for the installation of patented clevis, band, ring, expansion and trapeze types as well as fabricated hangers and supports made from structural angles, channels, etc.

In many cases, the drawings will not show hangers and supports but the specifications will state that they are to be furnished and installed by the contractor. Thus, this becomes the estimator's problem for the purpose of bidding the job. You will find under Section Ten entitled "Technical Information" on pages 214 through 221, diagrams, tables, formulas and solutions as to how a process piping system should be hung and/or supported.

HANGERS AND SUPPORTS

Fabrication: Labor only for fabrication of other than standard manufactured hangers and supports can be performed at 0.07 man hours per pound.

Field Erection:

Type of Hanger	Hanger Suspended From	Man Hours Per Hanger* Hanger Fastened To			
		Steel	Concrete or Masonry	Wood	Existing Pipe
PATENT Clevis Hanger Band Hanger Ring Hanger Expansion Hanger	Welded Clip Angle	1.50	--	--	--
	Clip Angle — Ramset	1.00	--	--	--
	Female Stud or Male Stud & Coupling — Ramset	.60	.60	--	--
	Female Stud or Male Stud & Coupling — Nelson Stud Welder	.60	--	--	--
	Beam Clamp or Corn Clamp	1.30	--	--	--
	Cinch Anchor	--	2.00	--	--
	Bolt or Strap	--	--	1.60	--
	Band and Rod	--	--	--	1.00
PATENT Trapeze Hanger (1' - 4' Bar)	Welded Clip Angles	2.00	--	--	--
	Clip Angle — Ramset	1.50	--	--	--
	Female Stud or Male Stud & Coupling — Ramset	1.20	--	--	--
	Female Stud or Male Stud & Coupling — Nelson Stud Welder	1.20	--	--	--
	Beam Clamp or Corn Clamp	2.00	--	--	--
	Cinch Anchor	--	4.00	--	--

*The patent hanger allowances are for supporting pipe through 4" size.

Fabricated Hangers (Angles, Channels, Etc.): 0.08 man hours per pound with a minimum time of 1 man hour regardless of weight.

The following factors should be applied for sizes over 4":

    6" — 1.20 man hours
    8" — 1.50 man hours
    10" — 1.80 man hours
    12" — 2.20 man hours

Section Seven

PAINTING

This section deals solely with the sandblasting and painting of a piping system and is so arranged as to include the direct man hours by pipe size for six (6) different types or specifications.

We have not covered color coding under this section due to the fact that the scope of the work involved in this operation can vary so greatly. As an example, you may be able to set-up in one location and band as many as a dozen lines, on the other hand the same set-up may be required to band one line. Therefore, we feel that this operation must be looked at individually according to piping specifications and locations.

SURFACE AREA OF PIPE FOR PAINTING

Nominal Size Inches	Surface Area S.F. Per L.F.	Nominal Size Inches	Surface Area S.F. Per L.F.
1	0.344	22	5.75
1-1/2	0.497	24	6.28
2	0.622	26	6.81
2-1/2	0.753	28	7.32
3	0.916	30	7.85
3-1/2	1.047	32	8.38
4	1.178	34	8.89
5	1.456	36	9.42
6	1.734	38	9.96
8	2.258	40	10.46
10	2.810	42	11.00
12	3.142	44	11.52
14	3.67	46	12.03
16	4.19	48	12.57
18	4.71	54	14.13
20	5.24	60	15.71

SAND BLAST AND PAINT PIPE

COMMERCIAL BLAST

NET MAN HOURS PER LINEAL FOOT

Nominal Size Inches	4-Coats Conventional Paint	4-Coats Chlorinated Rubber	4-Coats Vinyl Paint	1-Coat Dimetcote #3	5-Coats Epoxy Paint	1/16" Barretts 10-70
2	0.05	0.05	0.06	0.05	0.08	0.04
2-1/2	0.05	0.06	0.08	0.07	0.10	0.05
3	0.06	0.07	0.09	0.08	0.12	0.06
3-1/2	0.07	0.08	0.10	0.09	0.13	0.07
4	0.08	0.08	0.10	0.10	0.14	0.07
5	0.09	0.10	0.13	0.11	0.17	0.08
6	0.10	0.12	0.15	0.13	0.19	0.10
8	0.13	0.15	0.19	0.17	0.24	0.13
10	0.16	0.18	0.23	0.20	0.29	0.15
12	0.18	0.19	0.25	0.21	0.32	0.16
14	0.19	0.21	0.27	0.24	0.35	0.18
16	0.22	0.24	0.31	0.27	0.40	0.21
18	0.25	0.27	0.35	0.31	0.45	0.24
20	0.28	0.31	0.40	0.34	0.50	0.26
22	0.31	0.34	0.44	0.38	0.55	0.29
24	0.34	0.37	0.47	0.42	0.60	0.32
26	0.37	0.40	0.52	0.46	0.65	0.35
28	0.40	0.43	0.56	0.49	0.70	0.37
30	0.42	0.46	0.59	0.52	0.75	0.40
32	0.45	0.49	0.64	0.56	0.80	0.43
34	0.48	0.52	0.68	0.60	0.85	0.45
36	0.50	0.56	0.71	0.63	0.90	0.48
38	0.53	0.59	0.76	0.67	0.95	0.51
40	0.56	0.62	0.80	0.70	1.00	0.53
42	0.59	0.65	0.83	0.74	1.06	0.56
44	0.62	0.68	0.88	0.77	1.10	0.59
46	0.64	0.71	0.92	0.81	1.15	0.61
48	0.65	0.72	0.96	0.84	1.20	0.64
54	0.73	0.87	1.08	0.95	1.35	0.72
60	0.85	0.93	1.19	1.05	1.51	0.80

Man hours for painting pipe only. Labor for scaffolding must be added.

Man hours for galvanizing exterior of pipe only is approximately 80% of conventional paint.

Man hours to galvanize exterior and interior of pipe is approximately the same as dimetcote.

Section Eight

PATENT SCAFFOLDING

This section covers labor in man hours for the erection and dismantling of patent tubular steel type scaffolding.

In the process of making the piping material take-off, the estimator should give due consideration to the lengths of run, the height, etc., so that the number and height of sections of scaffolding may be determined for the entire piping job.

We have not attempted to cover job fabricated homemade scaffolding due to the fact that this type of scaffolding for a piping job is so outrageously high. If this type of scaffolding is desired, you must look elsewhere or draw from your past experience.

ERECT AND DISMANTLE

DIRECT LABOR — MAN HOURS PER SECTION

Patent Tubular Steel Scaffolding — 2" Planking Top.
Sections — 7' L x 5' W x 5' H

Includes: Transporting scaffolding and materials from storage.
Erection of scaffolding including leveling and securing.
Installation of 2" planking.
Dismantling of scaffolding.
Transporting scaffolding and materials to storage.

	MAN HOURS PER SECTION					
	One or Two Sections High			More than Two Sections High		
	Erect	Dismantle	Total	Erect	Dismantle	Total
One to two sections long	1.40	1.00	2.40	1.70	1.20	2.90
Three to five sections long	0.90	0.60	1.50	1.00	0.70	1.70
Six sections and more long	0.70	0.40	1.10	0.90	0.50	1.40

Section Nine

INSULATION

The hardest of all piping items for which to try to set a standard man hour rate is insulation. This is due largely to the fact that this is a very special item which is usually subcontracted to an organization who specializes in this field. Too, an insulation contractor will consider many factors before he submits his bid — such as, "Do I want or need this job, is the job large or small, etc." The cost of moving in and setting up is just as great regardless of the size of the job.

The man hours which appear in the following tables are the average of many jobs and we believe they will work fine for the types of insulation they cover. However, we believe that for projects where much and varied insulation is to be used a contractor who specializes in this type of work should be consulted on this matter.

INDOOR THERMAL TYPE
NET MAN HOURS

Thick-ness Inches	Pipe Size	Straight Pipe per LF	Bent Pipe per LF	Flanges Line per Ea.	Valves Flgd. per Ea.	Valves. S & W per Ea.	Fittings Flanged per Ea.	Fittings S & W per Ea.	Hangers Pipe per Ea.	Nozzles per Each
1.0	1/2	.18	.28	.56	1.50	.75	1.50	.28	.18	.18
	3/4	.19	.29	.59	1.58	.79	1.58	.29	.19	.19
	1	.21	.31	.63	1.69	.84	1.69	.31	.21	.21
	1-1/2	.24	.36	.72	1.92	.96	1.92	.36	.24	.24
	2	.25	.38	.76	2.04	1.02	2.04	.38	.25	.25
	3	.31	.47	.94	2.52	1.26	2.52	.47	.31	.31
	4	.37	.56	1.12	2.99	1.49	2.99	.74	.37	.37
	6	.43	.64	1.29	3.45	1.72	3.45	.86	.43	.43
1.5	1/2	.28	.43	.86	2.30	1.15	2.30	.43	.28	.28
	3/4	.30	.45	.90	2.42	1.21	2.42	.45	.30	.30
	1	.31	.47	.95	2.54	1.27	2.54	.47	.31	.31
	1-1/2	.35	.53	1.06	2.84	1.42	2.84	.53	.35	.35
	2	.37	.56	1.13	3.01	1.50	3.01	.56	.37	.37
	3	.44	.66	1.34	3.57	1.78	3.57	.66	.44	.44
	4	.50	.76	1.52	4.06	2.03	4.06	1.01	.50	.50
	6	.57	.86	1.73	4.63	2.31	4.63	1.15	.57	.57
	8	.67	1.01	2.03	5.43	2.71	5.43	1.69	.67	.67
	10	.80	1.21	2.43	6.48	3.24	6.48	2.02	.80	.80
	12	.91	1.36	2.73	7.30	3.65	7.30	2.73	.91	.91
	14	1.01	1.52	3.05	8.14	4.07	8.14	3.05	1.01	1.01
	16	1.14	1.71	3.43	9.15	4.57	9.15	4.56	1.14	1.14
	18	1.27	1.90	3.80	10.17	5.08	10.17	6.35	1.27	1.27
	20	1.39	2.08	4.17	11.13	5.56	11.13	6.94	1.39	1.39
	24	1.62	2.43	4.87	12.99	6.49	12.99	9.74	1.62	1.62
2.5	1/2	.47	.71	1.42	3.79	1.89	3.79	.71	.47	.47
	3/4	.48	.72	1.45	3.88	1.94	3.88	.72	.48	.48
	1	.50	.76	1.52	4.06	2.03	4.06	.76	.50	.50
	1-1/2	.55	.82	1.65	4.41	2.20	4.41	.82	.55	.55
	2	.58	.87	1.74	4.65	2.32	4.65	.87	.58	.58
	3	.68	1.02	2.04	5.30	2.72	5.30	1.02	.68	.68
	4	.78	1.16	2.33	6.21	3.10	6.21	1.55	.78	.78
	6	.86	1.28	2.58	6.88	3.44	6.88	1.72	.86	.86
	8	.97	1.46	2.93	7.81	3.90	7.81	2.43	.97	.97
3.5	1/2	.74	1.12	2.24	6.00	3.00	6.00	1.12	.74	.74
	3/4	.78	1.18	2.36	6.31	3.15	6.31	1.18	.78	.78
	1	.80	1.20	2.40	6.42	3.21	6.42	1.20	.80	.80
	1-1/2	.86	1.29	2.59	6.91	3.45	6.91	1.29	.86	.86
	2	.91	1.37	2.74	7.32	3.66	7.32	1.37	.91	.91
	3	1.02	1.54	3.08	8.22	4.11	8.22	1.54	1.02	1.02
	4	1.11	1.67	3.34	8.93	4.46	8.93	2.23	1.11	1.11
	6	1.21	1.81	3.63	9.69	4.84	9.69	2.42	1.21	1.21
	8	1.35	2.03	4.06	10.84	5.42	10.84	3.38	1.35	1.35

Thermal Insulation: Consists of applying hydraulic setting, insulating cement by spraying, brushing, troweling or palming, coating with vinyl emulsion, double wrapping with glass fiber cloth and coating with vinyl emulsion seal coat.

Outside Use: Add 10% to above man hours.

Foamglass: Use same man hours as appear above for this type insulation. This will include labor for butter joints with "Seal Koat" and secure with 16 and 14 gauge galvanized wire on 9" centers. Finish with one coat "Seal Koat" for indoor piping and 55# asbestos roofing felt secured with 16 gauge wire 6" on center over the layer of "Seal Koat" on outside piping.

Note: S & W denotes screwed and welded.

INSULATION

HOT PIPING — MAN HOURS

Pipe Size Inches	Thickness and Type	Straight Pipe per l. f.	Screwed & Weld Fittings per each	Flanges per pair	Flanged Valves & Fittings each
1/2	1" thick Calsilite	.11	.14	.36	.74
3/4	1" thick Calsilite	.11	.15	.36	.74
1	1" thick Calsilite	.12	.18	.36	.74
1-1/2	1" thick Calsilite	.13	.21	.41	.83
2	1" thick Calsilite	.14	.22	.44	.88
3	1" thick Calsilite	.18	.27	.54	1.37
4	1" thick Calsilite	.21	.34	.65	1.65
5	1" thick Calsilite	.25	.52	.72	2.05
6	1" thick Calsilite	.25	.61	.77	2.15
7	1-1/2" thick Calsilite	.33	.93	.96	2.96
8	1-1/2" thick Calsilite	.36	1.18	1.10	3.39

Man Hour:

1. Above thicknesses and man hours are for all hot services, if calcium silicate is used.
2. The above man hours are for either *indoor* or *outdoor* service.
3. *Bent Pipe:* 1.5 x straight pipe of like size and thickness measured along outside radius.
4. *Steam Traced Piping:* To be man houred at size of pipe covering required to fit over pipe and tracer line.
5. *Method of Measurement:* Straight pipe to be determined by measuring along approximate center line over the exterior of the insulation from center line to center line of change of direction. Measurement shall be made through all valves and fittings, except bent pipe.
6. *Specifications:*
   a. *Pipe Covering:* Molded sections secured with 16 ga. galvanized tie wire. *Finish:* Indoors with 6 ounce canvas with laps sealed with Arabol lagging adhesive. *Finish:* Outdoors with 55# Fiberock Asbestos Roofing felt secured with 16 ga. galvanized tie wire 6" o. c.
   b. *Fittings:* To be built-up with insulating cement or sectional pipe covering pointed up with asbestos cement, finished with 6 ounce canvas and Arabel for indoor service and "Seal Perm" for outdoor service.

Section Ten

SAMPLE ESTIMATE

This section is presented for the purpose of showing the work ability of a few of the man hour charts as appear throughout this manual. It does not mean that a take-off must be made in this manner before the man hour charts will work. It is merely a suggested method.

You will note on the following take-off sheets at the top of the page a predetermined composite rate, arrived at as outlined in the Introduction of this manual. Simply by multiplying this composite rate by the total man hours involved, a total estimated direct labor dollar value can be easily and accurately obtained.

We do not show in this sample estimate any material cost, nevertheless, you will find ample space provided for this item. You will also find space provided for both unit and total weights of pipe and fittings. We feel that this item has much value such as an estimate check using the weight method, or for the estimation of warehousing, equipment usage and fabrication shop set-up.

We purposely have not included material, miscellaneous supplies, equipment usage, overhead and profit in this estimate. As is stated in the Preface of this manual, its sole purpose is for the estimation of direct labor in man hours only.

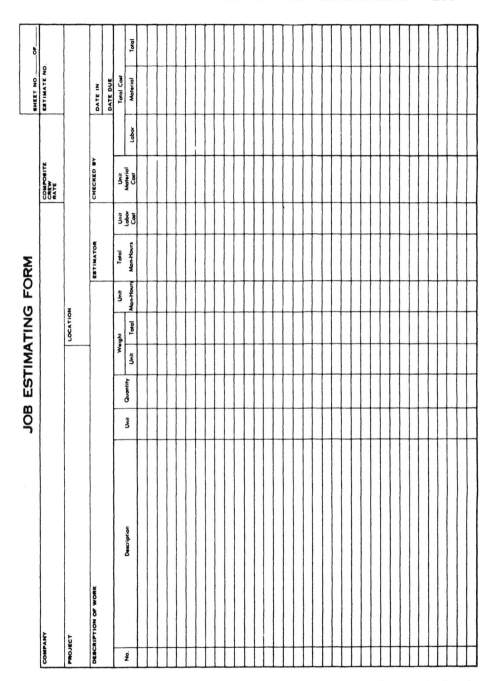

This Job Estimating Form is ideal for use when working with the Estimating Man-Hour Manuals.

JOB ESTIMATING FORM

SHEET NO 1 OF 7
ESTIMATE NO 203

COMPANY	COMPOSITE CREW RATE
AMERICAN CHEMICAL COMPANY	11.11

PROJECT: PROCESS PIPING
LOCATION: ANYWHERE, U.S.A.

DESCRIPTION OF WORK: SHOP FABRICATION – CARBON STEEL – Owner Furnished Materials

ESTIMATOR: Page
CHECKED BY: Nation
DATE IN 7-3-75
DATE DUE 7-5-75

No	Description	Unit	Quantity	Weight Unit	Weight Total	Unit Man-Hours	Total Man-Hours	Unit Labor Cost	Unit Material Cost	Total Cost Labor	Total Cost Material
1	3" – 90° Sch. 40 Wld. Ell	Pcs.	10	4.6	46	Hdlg.	Hdlg.		Owner	Owner	
2	4" – " " " " "	"	8	8.7	70	Incl.	Incl.		Furn.	Furn.	
3	6" – " " " " "	"	3	23.0	69	w/Pipe	w/Pipe		"	"	
4	3" – 45° " " " "	"	5	2.3	12				"	"	
5	4" – Sch. 40 Wld. Tee	"	3	12.6	38				"	"	
6	4" x 3" – Sch. 40 Wld. Red	"	2	3.6	7				"	"	
7	3" – Sch. 40 Wld. Cap	"	2	1.4	3				"	"	
8	4" – Sch. 40 Butt Welds	Ea.	32	—	—	1.3	41.6		"	462	"
9	3" – " " " "	"	50	—	—	1.1	55.0		"	611	"
10	6" – Sch. 80 " "	"	15	—	—	2.1	31.5		"	350	"
11	4" – Sch. 40 Cut	"	2	—	—	.2	.4		"	4	"
12	4" – " " U-Bevel	"	2	—	—	1.5	3.0		"	33	"
13	3" – 150# S.O. Flgs.	Pcs.	2	10	20	1.6	3.2		"	36	"
14	3" – Sch. 40 45° Bend	"	2	—	—	3.1	6.2		"	69	"
15	4" – Sch. 40 90° Bend	"	1	—	—	4.4	4.4		"	49	"
16	6" – Sch. 80 Offset Bend	"	1	—	—	10.1	10.1		"	112	"
17	6" – Sch. 80 Preheat	Ea.	1	—	—	.9	.9		"	10	"
18	6" – Sch. 80 Stress Relieve	"	1	—	—	3.6	3.6		"	40	"
19	3" – Test Fab. Assy. (2-Outlets)	"	1	—	—	5.2	5.2		"	58	"
20	3" – Sch. 40 Smls A-53 Pipe	LF	200	7.6	1520	.041	8.2		"	91	"
21	4" – " " " "	"	350	10.8	3780	.045	15.8		"	176	"
22	6" – " " " "	"	100	28.6	2860	.070	7.0		"	78	"
	TOTAL FITTING WEIGHT 265						TOTAL THIS SHEET			2179	
	PIPE 8160										

JOB ESTIMATING FORM

COMPANY: AMERICAN CHEMICAL COMPANY

PROJECT: PROCESS PIPING
LOCATION: ANYWHERE, USA

DESCRIPTION OF WORK: SHOP FABRICATION – ALLOY – Owner Furnished Materials

ESTIMATOR: Page
CHECKED BY: Nation

COMPOS/CREW RATE: 11.11

SHEET NO 2 OF 7
ESTIMATE NO 203
DATE IN 7-3-75
DATE DUE 7-5-75

No	Description	Unit	Quantity	Weight Unit	Weight Total	Unit Man-Hours	Total Man-Hours	Unit Labor Cost	Unit Material Cost	Labor	Material	Total
1	8" - 90° Sch. 40 Wld. Ell	Pcs	8	46.0	368	Hdlg.	Hdlg.		Owner		Owner	
2	6" - " " " "	"	6	23.0	138	Incl	Incl		Furn.		Furn	
3	8" - 45° " " "	"	2	23.0	46	w/pipe	w/pipe					
4	8" - 300# W.N. Flgs.	"	10	67.0	670	3.83	38.30		"	426	"	
5	6" - " " "	"	8	42.0	336	2.97	23.76		"	264	"	
6	8" - Sch. 40 Butt Welds	Ea.	35	—	—	3.83	134.05		"	1489	"	
7	6" - " " "	"	25	—	—	2.97	74.25		"	825	"	
8	8" S.40 Snls. 18-8 Type 304 Pipe	LF	180	28.55	5367	.063	11.34		"	126	"	
9	6" " " " " "	"	60	18.97	1138	.051	3.06		"	34	"	
	FIELD ERECT - SHOP FABRICATED PIPING											
1	3" S.40 Shop Fab Spool 10' long	Pcs.	20	76.0	1520	3.9	78.00		"	867	"	
2	4" " " " "	"	35	108.0	3780	4.1	143.50		"	1594	"	
3	6" " " "	"	6	190.0	1140	4.7	28.20		"	313	"	
4	6" S.80 " "	"	10	286.0	2860	6.4	64.00		"	711	"	
5	8" S.40 " "	"	18	286.0	5148	5.7	102.60		"	1140	"	
6	3" Bolt-Ups 150#	Ea.	40	1.5	60	.8	32.00		"	356	"	
7	4" " "	"	70	4.0	280	1.2	84.00		"	933	"	
8	6" " " 300#	"	32	11.5	368	1.7	54.40		"	604	"	
9	8" " " "	"	36	18.0	648	2.4	86.40		"	960	"	

TOTAL FITTING WT. 2914
PIPE " 20953
TOTAL THIS SHEET 10642.00

JOB ESTIMATING FORM

SHEET NO 2 OF 7
ESTIMATE NO 203

COMPANY: AMERICAN CHEMICAL COMPANY
PROJECT: PROCESS PIPING
LOCATION: ANYWHERE, USA
DESCRIPTION OF WORK: FIELD FABRICATE AND ERECT—SCREWED - Owner Furnished Materials

COMPOSITE CREW RATE 11.11
ESTIMATOR Page
CHECKED BY Nation
DATE IN 7-3-75
DATE DUE 7-5-75

No	Description	Unit	Quantity	Weight Unit	Weight Total	Man-Hours Unit	Man-Hours Total	Unit Labor Cost	Unit Material Cost	Total Cost Labor	Total Cost Material	Total
1	1/2" 90° Ell Scrd. 150# M.I.	Pcs.	15	.25	4	.2	3.0		Owner	33	Owner	
2	1" " " " "	"	10	.60	6	.4	4.0		Furn.	44	Furn.	
3	2" " 2000# f.s.	"	5	4.00	20	.6	3.0		"	33	"	
4	1/2" 45° Ell Scrd. 150# M.I.	"	4	.23	1	.2	.80		"	9	"	
5	1" " " " "	"	2	.52	1	.4	.80		"	9	"	
6	2" " 2000# f.s.	"	1	3.00	3	.6	.6		"	7	"	
7	1" Tee Scrd. 150 # M.I.	"	2	.86	2	.6	1.2		"	13	"	
8	2" " Scrd. 2000# f.s.	"	1	5.00	5	.9	.9		"	10	"	
9	1" Cross Scrd. 150# M.I.	"	1	.97	1	.8	.8		"	9	"	
10	2" " 2000# f.s.	"	1	5.00	5	1.2	1.2		"	13	"	
11	1" x 1/2" Swgs. S.40 T.B.E.	"	2	1.00	2	.3	.6		"	7	"	
12	2" x 1" Swgs. S. 80 T.B.E.	"	1	2.00	2	.5	.5		"	6	"	
13	2" Coup. 2000# f.s.	"	2	1.05	2	.6	1.2		"	13	"	
14	1/2" Union 150# M.I.	"	10	.38	4	.2	2.0		"	22	"	
15	1" " " "	"	5	.90	5	.4	2.0		"	22	"	
16	2" " 2000# f.s.	"	3	5.00	15	.6	1.8		"	20	"	
17	2" - 150# Scrd. Flgs.	"	2	5.00	10	1.2	2.4		"	27	"	
18	1" x 1/2" Scrd. Red. 150# M.I.	"	2	.44	1	.3	.6		"	7	"	
19	2" x 1" " 2000# f.s.	"	1	3.00	3	.5	.5		"	6	"	
20	1/2" x 6" Nipple Sch. 40	"	5	.38	2	.2	1.0		"	11	"	
21	1" x 6" " 80	"	2	.99	2	.4	.8		"	9	"	
22	1/2" Sch. 40 Buttweld T&C pipe	LF	100	.85	85	.16	16.0		"	178	"	
23	1" " " " "	"	75	1.68	126	.17	12.75		"	142	"	
24	1" " 80	"	50	5.02	251	.24	12.0		"	133	"	
25	2" - 150# Bolts & Gaskets	Sets	2	1.50	3	.7	1.4		"	16	"	
26	1/2" Make-Ons	Ea.	5	--	--	.1	.5		"	6	"	
27	1" " "	"	4	--	--	.2	.8		"	9	"	
28	2" " "	"	3	--	--	.3	.9		"	10	"	
	TOTAL FITTING WEIGHT				99							
	TOTAL PIPE WEIGHT				462			TOTAL THIS SHEET		824.00		

JOB ESTIMATING FORM

SHEET NO 4 OF 7
ESTIMATE NO 203

COMPANY: AMERICAN CHEMICAL COMPANY

PROJECT: PROCESS PIPING

LOCATION: ANYWHERE, USA

COMPOSITE CREW RATE 11.11

DESCRIPTION OF WORK: FIELD FABRICATE AND ERECT – WELDED – Owner Furnished Materials

ESTIMATOR: Page

CHECKED BY: Nation

DATE IN 7-3-75

DATE DUE 7-5-75

No	Description	Unit	Quantity	Weight Unit	Weight Total	Unit Man-Hours	Total Man-Hours	Unit Labor Cost	Unit Material Cost	Total Cost Labor	Total Cost Material	Total Cost Total
1	3" Sch. 40 90° Wld. Ell.	Pcs.	8	4.6	37	Hdlg. Incl.	Hdlg. Incl.		Owner Furn.		Owner Furn.	
2	4" " 80 " "	"	15	8.7	131	w/pipe	w/pipe		Furn.		Furn.	
3	6" " 80 " "	"	5	34.0	170	"	"		"		"	
4	4" Sch. 40 45° Wld. Ell.	"	4	4.3	17	"	"		"		"	
5	6" Sch. 80 Wld. Tee	"	2	42.0	84	"	"		"		"	
6	4" x 3" Sch. 40 Wld. Red	"	1	3.6	4	"	"		"		"	
7	6" Sch. 80 Wld. Cap	"	1	9.2	9	"	"		"		"	
8	3" - 150# Bolts & Gaskets	Sets	7	1.5	11	.8	5.6		"	62	"	
9	4" - 300#	"	10	7.5	75	1.4	14.0		"	156	"	
10	6" - 600#	"	2	30.0	60	1.8	3.6		"	40	"	
11	3" - Sch. 40 Butt-welds	Ea.	22	—	—	1.3	28.60		"	318	"	
12	4" - " "	"	50	—	—	1.5	75.00		"	833	"	
13	6" - " 80 " "	"	20	—	—	2.5	50.00		"	556	"	
14	4" Sch. 40 90° Nozzle Weld	"	2	—	—	4.0	8.00		"	89	"	
15	3" Sch. 40 Mitre Weld	"	1	—	—	1.95	1.95		"	22	"	
16	3" - 150# Flg. S.O.	Pcs.	2	10.0	20	1.8	3.60		"	40	"	
17	6" - 600#	"	1	95.0	95	5.9	5.9		"	66	"	
18	3" - 150# Flg. W.N.	"	3	10.0	30	1.3	3.9		"	43	"	
19	4" - 300#	"	10	25.0	250	1.5	15.0		"	167	"	
20	6" - 600#	"	1	85.0	85	2.0	2.0		"	22	"	
21	3" - 150# Flg. Bld.	"	2	11.0	22	.8	1.6		"	18	"	
22	3" Sch. 40 Smls. A-53 Pipe	LF	100	7.57	757	.23	23.0		"	256	"	
23	4" " " " A-106 Pipe	"	230	10.79	2482	.25	57.50		"	639	"	
24	6" Sch. 80 " "	"	50	28.57	1429	.38	19.00		"	211	"	
	TOTAL FITTING WEIGHT				1100		TOTAL THIS SHEET			3538.00		
	TOTAL PIPE WT.				4670							

JOB ESTIMATING FORM

SHEET NO. 5 of 7
ESTIMATE NO 203

COMPANY									COMPOSITE CREW RATE	11.11
AMERICAN CHEMICAL COMPANY										

PROJECT: PROCESS PIPING
LOCATION: ANYWHERE, USA

DESCRIPTION OF WORK: ERECT VALVES – SCREWED AND FLANGED – Owner Furnished Materials

ESTIMATOR: Page
CHECKED BY: Nation
DATE IN: 7-3-75
DATE DUE: 7-5-75

No	Description	Unit	Quantity	Weight Unit	Weight Total	Unit Man-Hours	Total Man-Hours	Unit Labor Cost	Unit Material Cost	Total Cost Labor	Total Cost Material	Total
1	1/2" Gate Va. Scrd.	Pcs.	20	2	40	.2	4.0		Owner	44	Owner	
2	1" " " "	"	15	4	60	.4	6.0		Furn.	67	Furn.	
3	2" " " "	"	10	12	120	.6	6.0		"	67	"	
4	1/2" Globe Va. Screwed	"	5	2	10	.2	1.0		"	11	"	
5	1" " " "	"	3	4	12	.4	1.2		"	13	"	
6	1" Check Valve Screwed	"	2	4	8	.4	.8		"	9	"	
7	2" " " "	"	1	12	12	.6	.6		"	7	"	
8	3" Gate Va. 150# Fldg.	"	10	97	970	2.8	28.0		"	311	"	
9	4" " 300# "	"	6	225	1350	4.8	28.8		"	320	"	
10	6" " " "	"	2	457	914	6.1	12.2		"	136	"	
11	6" " 600# "	"	1	743	743	6.9	6.9		"	77	"	
12	8" " 300# "	"	2	683	1366	8.2	16.4		"	182	"	
13	3" Globe Valve 150# Fldg.	"	1	100	100	2.8	2.8		"	31	"	
14	4" " 300# Fldg.	"	1	220	220	4.8	4.8		"	53	"	
15	3" Check Valve 150# "	"	1	71	71	2.8	2.8		"	31	"	
16	3" Plug Va. 150# Flgd.	"	1	42	42	2.8	2.8		"	31	"	
	TOTAL VALVE WT.				5295		TOTAL THIS SHEET			1390.00		

JOB ESTIMATING FORM

SHEET NO 6 OF 7
ESTIMATE NO 203

COMPANY: AMERICAN CHEMICAL COMPANY
PROJECT: PROCESS PIPING
DESCRIPTION OF WORK: HANGERS AND SUPPORTS - Owner Furnished Materials
LOCATION: ANYWHERE, USA
COMPOSITE CREW LABOR RATE: Fitter 11.11
ESTIMATOR: Page
CHECKED BY: Nation
DATE IN 7-3-75
DATE DUE 7-5-75

No	Description	Unit	Quantity	Weight Unit	Weight Total	Unit Man-Hours	Total Man-Hours	Unit Labor Cost	Unit Material Cost	Total Cost Labor	Total Cost Material	Total
1	Ring Hanger - WH Clip Angle & Steel	Ea.	25	—	—	1.50	37.50		Owner Furn.	417	Owner Furn.	
2	Trapeze Hanger - Wld. Clip Angle	"	15	—	—	2.00	30.00		"	333	"	
3	Fabricated Hangers	Lbs.	65	—	650	.08	52.00		"	578	"	
	SAND-BLAST & PAINT PIPE											
1	3" Pipe - 4 coats Conv. Paint	LF	300	—	—	.06	18.0		"	200	"	
2	4" " " " "	"	580	—	—	.08	46.40		"	516	"	
3	6" " 4 coats Vinyl Paint	"	150	—	—	.15	22.50		"	250	"	
	INSULATION											
1	8" Pipe - 1-1/2" Thk. Calsite	LF	180	—	—	.36	64.80		"	720	"	
	HAND EXCAVATE											
1	Medium - Gem Dry - First Lift	CY	20	—	—	1.60	32.00		"	160	"	
	Back Filling & Tamping											
1	Medium Soil	"	15	—	—	.80	12.0		"	60	"	
	Underground Piping											
1	6" - 150# B & S C.I. Pipe	LF	200	—	—	.11	22.0		"	244	"	
2	6" - 1/4 Bends (Lead Joint)	Ea.	10	—	—	.57	5.7		"	63	"	
3	6" - 1/8 Bends (Lead Joint)	"	5	—	—	.57	2.85		"	32	"	
4	6" Make-Ons (Lead Joint)	"	40	—	650	.57	22.80		"	253	"	

TOTAL HANGER WT. 650

TOTAL THIS SHEET 3826.00

JOB ESTIMATING FORM

COMPANY
AMERICAN CHEMICA L COMPANY

PROJECT
PROCESS PIPING

LOCATION
ANYWHERE, USA

COMPOSITE CREW RATE

SHEET NO. 7 OF 7

ESTIMATE NO. 203

CHECKED BY
Nation

ESTIMATOR
Page

DATE IN 7-3-75
DATE DUE 7-5-75

DESCRIPTION OF WORK
SUMMARY

No.	Description	Unit	Quantity	Weight Unit	Weight Total	Unit Man-Hours	Total Man-Hours	Unit Labor Cost	Unit Material Cost	Total Cost Labor	Total Cost Material	Total
1	Shop fabrication - Carbon Steel								2179.00			
2	Shop Fabrication - Alloy											
	and Field Erect - Shop Fabricated Piping								10642.00			
3	Field Fabricate and Erect - Screwed								824.00			
4	Field Fabricate and Erect - Welded								3538.00			
5	Erect Valves - Screwed and Flanged								1390.00			
6	Hangers and Supports, Sandblast and Paint Pipe											
	Insulation, Hand Exca vate, and Underground Piping								3826.00			
	TOTAL DIRECT LABOR								22399.00			

Section Eleven

TECHNICAL INFORMATION

As we stated in the Preface of this book, its intention is solely for the estimation of labor and is not intended for the design of piping. Therefore, this section has been held to a minimum and includes only information that we feel will benefit the estimator in the preparation of his estimate.

Included in this section are tables showing the circumferences of pipes for welding purposes, the amount of materials needed for insulation, the weights of pipe, fittings and valves and methods of hanging and supporting pipe and fittings.

We wish to acknowledge and to express our appreciation to the Grinnell Company, Inc., of Providence, Rhode Island, who has so graciously allowed us to reproduce the following tables.

CIRCUMFERENCES OF PIPE FOR COMPUTING WELDING MATERIAL

CIRCUMFERENCE OF PIPE IN INCHES

Nominal Pipe Size	Schedule Numbers									
	10	20	30	40	60	80	100	120	140	160
1	--	--	--	3.98	--	4.27	--	--	--	4.71
1-1/4	--	--	--	4.81	--	5.13	--	--	--	5.50
1-1/2	--	--	--	5.62	--	5.97	--	--	--	6.47
2	--	--	--	7.25	--	7.65	--	--	--	8.44
3	--	--	--	10.78	--	11.31	--	--	--	12.17
4	--	--	--	14.06	--	14.68	--	15.31	--	15.90
5	--	--	--	17.33	--	18.06	--	18.85	--	19.64
6	--	--	--	20.61	--	21.56	--	22.38	--	22.44
8	--	26.70	26.87	27.16	27.68	28.27	28.86	29.64	30.23	30.83
10	--	32.99	33.34	33.71	34.56	35.14	35.93	36.71	37.70	38.48
12	--	39.27	39.77	40.25	41.23	42.02	43.00	43.98	44.77	45.94
14	45.55	45.94	46.34	46.73	47.71	48.69	49.87	50.85	51.84	52.82
16	51.84	52.23	52.62	53.41	53.82	55.56	56.74	57.92	59.29	60.27
18	58.12	58.51	59.29	60.09	61.26	62.44	63.81	65.19	66.36	67.74
20	64.40	65.19	65.97	66.56	67.93	69.31	70.88	72.26	73.83	75.20
24	76.97	77.75	78.54	79.71	81.48	83.05	85.02	86.78	88.35	90.12

CIRCUMFERENCES OF HEAVY WALL PIPE
FOR COMPUTING WELDING MATERIAL

CIRCUMFERENCE OF PIPE IN INCHES

Nominal Pipe Size	Wall Thickness in Inches							
	.500	.750	1.00	1.25	1.50	1.75	2.00	2.25
3	12.57	14.14	15.71	17.28	18.85	20.42	21.99	23.56
4	15.71	17.28	18.85	20.42	21.99	23.56	25.13	26.70
5	18.85	20.42	21.99	23.56	25.13	26.70	28.27	29.85
6	21.99	23.56	25.13	26.70	28.27	29.85	31.42	32.99
8	28.27	29.85	31.42	32.99	34.56	36.13	37.70	39.27
10	34.56	36.13	37.70	39.27	40.84	42.41	43.98	45.55
12	40.84	42.41	43.98	45.55	47.12	48.69	50.27	51.84
14	47.12	48.69	50.27	51.84	53.41	54.98	56.55	58.12
16	53.41	54.98	56.55	58.12	59.69	61.26	62.83	64.40
18	59.69	61.26	62.83	64.40	65.97	67.54	69.12	70.69
20	65.97	67.54	69.12	70.69	72.26	73.83	75.40	76.37
22	72.26	73.83	75.40	76.97	78.54	80.11	81.68	83.25
24	78.54	80.11	81.68	83.25	84.82	86.39	87.96	89.54
	2.50	2.75	3.00	3.25	3.50	3.75	4.00	4.25
10	47.12	48.69	50.27	51.84	53.41	54.98	56.55	58.12
12	53.41	54.98	56.55	58.12	59.69	61.26	62.83	64.40
14	59.69	61.26	62.83	64.40	65.97	67.54	69.12	70.69
16	65.97	67.54	69.12	70.69	72.26	73.83	75.40	76.97
18	72.26	73.83	75.40	76.97	78.54	80.11	81.68	83.25
20	78.54	80.11	81.68	83.25	84.82	86.39	87.96	89.54
22	84.82	86.39	87.96	89.54	91.11	92.68	94.25	95.82
24	91.11	92.68	94.25	95.82	97.39	98.96	100.53	102.10
	4.50	4.75	5.00	5.25	5.50	5.75	6.00	
20	91.11	92.68	94.25	95.82	97.39	98.96	100.53	
22	97.39	98.96	100.53	102.10	103.67	105.24	106.81	
24	103.67	105.24	106.81	108.39	109.96	111.53	113.10	

CIRCUMFERENCES OF LARGE O.D. PIPE FOR COMPUTING WELDING MATERIAL

CIRCUMFERENCE OF PIPE IN INCHES

Nominal Pipe Size	WALL THICKNESS IN INCHES							
	.375	.500	.750	1.00	1.25	1.50	1.75	2.00
26	84.04	84.82	86.39	87.96	89.54	91.11	92.68	94.25
28	90.32	91.11	92.68	94.25	95.82	97.39	98.96	100.53
30	96.60	97.39	98.96	100.53	102.10	103.67	105.24	106.81
32	102.89	103.67	105.24	106.81	108.39	109.96	111.53	113.10
34	109.17	109.96	111.53	113.10	114.67	116.24	117.81	119.38
36	115.45	116.24	117.81	119.38	120.95	122.52	124.09	125.66
38	121.74	122.52	124.09	125.66	127.23	128.81	130.38	131.95
40	128.02	128.81	130.38	131.94	133.52	135.09	136.66	138.23
42	134.30	135.09	136.66	138.23	139.80	141.37	142.94	144.51
44	140.59	141.37	142.94	144.51	146.08	147.66	149.23	150.80
46	146.87	147.66	149.23	150.80	152.36	153.94	155.51	157.08
48	153.15	153.94	155.51	157.08	158.65	160.22	161.79	163.36
54	172.00	172.79	174.36	175.93	177.50	179.07	180.64	182.21
60	190.82	191.64	193.21	194.78	196.35	197.92	199.49	201.06

	2.25	2.50	2.75	3.00	3.25	3.50	3.75	4.00
26	95.82	97.39	98.96	100.53	102.10	103.67	105.24	106.81
28	102.10	103.67	105.24	106.81	108.39	109.96	111.53	113.10
30	108.39	109.96	111.53	113.10	114.67	116.24	117.81	119.38
32	114.67	116.24	117.81	119.38	120.95	122.52	124.09	125.66
34	120.95	122.52	124.09	125.66	127.23	128.81	130.38	131.95
36	127.23	128.81	130.38	131.95	133.52	135.09	136.66	138.23

	4.25	4.50	4.75	5.00	5.25	5.50	5.75	6.00
26	108.39	109.96	111.53	113.10	114.67	116.24	117.81	119.38
28	114.67	116.24	117.81	119.38	120.95	122.52	124.09	125.66
30	120.95	122.52	124.09	125.64	127.23	128.81	130.38	131.95
32	127.23	128.81	130.38	131.95	133.52	135.09	136.66	138.23
34	133.52	135.09	136.66	138.23	139.80	141.37	142.94	144.51
36	139.80	141.37	142.94	144.51	146.08	147.66	149.23	150.80

WEIGHTS OF PIPING MATERIALS

The weight per foot of steel pipe is subject to the following tolerances:

SPECIFICATION	TOLERANCE
A.S.T.M. A-53 } A.S.T.M. A-120 }	STD WT + 5%, − 5% XS WT + 5%, − 5% XXS WT +10%, −10%
A.S.T.M. A-106	SCH 10–120 +6.5%, −3.5% SCH 140–160 +10%, −3.5%
A.S.T.M. A-158 } A.S.T.M. A-206 } A.S.T.M. A-280 }	12″ and under +6.5%, −3.5% over 12″ +10%, −5%
API 5L	All sizes +6.5%, −3.5%

Weight of Tube $= F \times 10.6802 \times T \times (D - T)$ pounds/foot

T = wall thickness in inches

D = outside diameter in inches

F = relative weight factor

The weight of tube furnished in this piping data is based on low carbon steel weighing 0.2833 pounds per cubic inch.

Relative Weight Factor F of various metals

Aluminum	= 0.35
Brass	= 1.12
Cast Iron	= 0.91
Copper	= 1.14
Lead	= 1.44
Ferritic Stainless Steel	= 0.95
Austenitic Stainless Steel	= 1.02
Steel	= 1.00
Tin	= 0.93
Wrought Iron	= 0.98

(Weight of Contents of a Tube $= G \times 0.3405 \times (D - 2T)^2$ pounds per foot)

G = Specific Gravity of Contents

T = Tube Wall Thickness, inches

D = Tube Outside Diameter, inches

The weight of Welding Tees and Laterals is for full size fittings. The weight of reducing fittings is approximately the same as for full size fittings.

The weight of Welding Reducers is for one size reduction, and is approximately correct for other reductions.

Pipe Covering temperature ranges are intended as a guide only and do not constitute a recommendation for specific thickness of material.

Pipe Covering thicknesses and weights indicate average conditions and include all allowance for wire, cement, canvas, bands, and paint. The listed thicknesses of combination covering is the sum of the inner and the outer layer thickness. When specific inner and outer layer thicknesses are known, add them, and use the weight for the nearest tabulated thickness.

To find the weight of covering on Fittings, Valves, or Flanges, multiply the weight factor (light faced subscript) by the weight per foot of covering used on straight pipe. All Flange weights include the proportional weight of bolts or studs required to make up all joints.

Lap Joint Flange weights include the weight of the lap.

Welding Neck Flange weights are compensated to allow for the weight of pipe displaced by the flange. Pipe should be measured from the face of the flange.

All Flanged Fitting weights include the proportional weight of bolts or studs required to make up all joints.

To find the approximate weight of Reducing Flanged Fittings, subtract the weight of a full size Slip-On Flange and add the weight of reduced size Slip-On Flange.

Weights of valves of the same type may vary because of individual Manufacturer's design. Listed valve weights are approximate only. When it is possible to obtain specific weights from the Manufacturer, such weights should be used.

To obtain the approximate weight of Flanged End Steel Valves, add the weight of two Slip-On Flanges of the same size and series to the weight of the corresponding Welding End Valves.

These pages reproduced through the courtesy of Grinnell Co., Inc.

WEIGHTS OF PIPING MATERIALS—1″ PIPE SIZE

PIPE

Schedule No.	40	80	160	
Wall Designation	Std.	XS		XXS
Thickness—In.	.133	.179	.250	.358
Pipe—Lbs/Ft	1.68	2.17	2.84	3.66
Water—Lbs/Ft	.37	.31	.23	.12

WELDING FITTINGS

L.R. 90° Elbow	.3 / .3	.4 / .3	.6 / .3	.7 / .3
S.R. 90° Elbow	.2 / .2			
L.R. 45° Elbow	.2 / .2	.3 / .2	.4 / .2	.4 / .2
Tee	.8 / .4	.9 / .4	1.1 / .4	1.3 / .4
Lateral				
Reducer	.3 / .2	.4 / .2	.4 / .2	.5 / .2
Cap	.2 / .3	.3 / .3	.4 / .3	.4 / .3

COVERING

	to 260°	260-360	360-440	440-525	525-600	600-700	700-800	800-900	900-1000
Temperature Range °F									
85% Magnesia — Thickness—In.	1/8	1/8	1 1/2	2	1 15/16				
85% Magnesia — Lbs/Ft	.65	.65	1.45	2.25	2.20				
Combination — Thickness—In.						2	2	2	2
Combination — Lbs/Ft						3.7	3.7	3.7	3.7
Calcium Silicate — Thickness In.	1	1	1	1	1 1/2	1 1/2	1 1/2	2	2
Calcium Silicate — Lbs/Ft	.75	.75	.75	.75	1.27	1.27	1.27	1.94	1.94

FLANGES

Pressure Rating psi	Cast Iron 125	Cast Iron 250	Steel 150	Steel 300	Steel 400	Steel 600	Steel 900	Steel 1500	Steel 2500
Screwed or Slip-On	2.5 / 1.5	4 / 1.5	2.5 / 1.5	4 / 1.5		5 / 1.5		12 / 1.5	15 / 1.5
Welding Neck			2.3 / 1.5	4.8 / 1.5		7 / 1.5		11 / 1.5	15 / 1.5
Lap Joint			2.5 / 1.5	4 / 1.5		5 / 1.5		12 / 1.5	15 / 1.5
Blind	2.5 / 1.5	4 / 1.5	2.5 / 1.5	5 / 1.5		5 / 1.5		12 / 1.5	15 / 1.5

FLANGED FITTINGS

	125	250	150	300	400	600	900	1500	2500
S.R. 90° Elbow	6 / 3.6					15 / 3.7		28 / 3.8	
L.R. 90° Elbow	8 / 3.8								
45° Elbow	5 / 3.2					14 / 3.4		26 / 3.6	
Tee	11 / 5.4					20 / 5.6		39 / 5.7	

**VALVES

	125	250	150	300	400	600	900	1500	2500
Flanged Bonnet Gate								67 / 4.3	
Flanged Bonnet Globe or Angle									
Flanged Bonnet Check									
Pressure Seal Bonnet—Gate							31 / 1.7	31 / 0.9	
Pressure Seal Bonnet—Globe									

Bolts

	125	250	150	300	400	600	900	1500	2500
*One Complete Flanged Joint	1	2	1	2		2		6	6

SEE GENERAL NOTES FOR MATERIALS NOT SHOWN

All weights are shown in bold type.

The weight of steel pipe is per linear foot.

For Boiler Feed Piping, add the weight of water to the weight of steel pipe.

The pipe covering thicknesses and weights indicate the average conditions per linear foot and include all allowances for wire, cement, canvas, bands and paint. The listed thickness of combination covering is the sum of the inner and the outer layer thickness.

Pipe covering temperature ranges are intended as a guide only and do not constitute a recommendation for specific thickness of materials.

To find the weight of covering on Flanged Fittings, Valves, or Flanges, multiply the weight factor (lightface subscript) by the weight per foot of covering used on straight pipe.

*All Flanged Fitting, Flanged Valve and Flange weights include the proportional weight of bolts or studs required to make up all joints.

**Cast Iron Valve weights are for flanged valves. Steel Valve weights are for welding end valves.

WEIGHTS OF PIPING MATERIALS—1¼″ PIPE SIZE

PIPE	Schedule No.	40	80	160							
	Wall Designation	Std.	XS		XXS						
	Thickness—In.	.140	.191	.250	.382						
	Pipe—Lbs/Ft	2.27	3.00	3.77	5.21						
	Water—Lbs/Ft	.65	.56	.46	.27						

WELDING FITTINGS	L.R. 90° Elbow	.6 / .3	.8 / .3	.9 / .3	1.2 / .3
	S.R. 90° Elbow	.4 / .2			
	L.R. 45° Elbow	.4 / .2	.5 / .2	.6 / .2	.8 / .2
	Tee	1.3 / .5	1.6 / .5	1.9 / .5	2.4 / .5
	Lateral	3.4 / 1.2	4.2 / 1.2		
	Reducer	.5 / .2	.5 / .2	.6 / .2	.8 / .2
	Cap	.3 / .3	.4 / .3	.5 / .3	.6 / .3

	Temperature Range °F	to 260°	260-360	360-440	440-525	525-600	600-700	700-800	800-900	900-1000	
COVERING	85% Magnesia Thickness—In.	¼	¼	1½	2	1¹⁵⁄₁₆					
	85% Magnesia Lbs/Ft	.80	.80	1.60	2.45	2.25					
	Combination Thickness—In.						2	2	2	2	
	Combination Lbs/Ft						4.1	4.1	4.1	4.1	
	Calcium Silicate Thickness—In.	1	1	1	1	1½	1½	2	2	2½	
	Calcium Silicate Lbs/Ft	.68	.68	.68	.68	1.19	1.19	1.87	1.87	2.68	

		Cast Iron		Steel							SEE GENERAL NOTES
	Pressure Rating psi	125	250	150	300	400	600	900	1500	2500	
FLANGES	Screwed or Slip-On	2.5 / 1.5	5 / 1.5	3.5 / 1.5	5 / 1.5		7 / 1.5		13 / 1.5	23 / 1.5	
	Welding Neck			3.3 / 1.5	6.5 / 1.5		7 / 1.5		12 / 1.5	23 / 1.5	
	Lap Joint			3.5 / 1.5	5 / 1.5		7 / 1.5		13 / 1.5	22 / 1.5	
	Blind	3.5 / 1.5	5 / 1.5	3.5 / 1.5	7 / 1.5		7 / 1.5		13 / 1.5	23 / 1.5	
FLANGED FITTINGS	S.R. 90° Elbow	8 / 3.6			17 / 3.7		18 / 3.8		33 / 3.9		
	L.R. 90° Elbow	10 / 3.9			13 / 3.9						
	45° Elbow	7 / 3.3			15 / 3.4		16 / 3.5		31 / 3.7		
	Tee	13 / 5.4			23 / 5.6		28 / 5.7		49 / 5.9		
VALVES	Flanged Bonnet Gate				34 / 3.8				95 / 4.4		
	Flanged Bonnet Globe or Angle										
	Flanged Bonnet Check				21 / 4						
	Pressure Seal Bonnet—Gate							38 / 1.8	38 / 1.1		
	Pressure Seal Bonnet—Globe										
Bolts	*One Complete Flanged Joint	1	2	1	2		2		6	9	

SEE GENERAL NOTES FOR MATERIALS NOT SHOWN

All weights are shown in bold type.

The weight of steel pipe is per linear foot.

For Boiler Feed Piping, add the weight of water to the weight of steel pipe.

The pipe covering thicknesses and weights indicate the average conditions per linear foot and include all allowances for wire, cement, canvas, bands and paint. The listed thickness of combination covering is the sum of the inner and the outer layer thickness.

Pipe covering temperature ranges are intended as a guide only and do not constitute a recommendation for specific thickness of materials.

To find the weight of covering on Flanged Fittings, Valves, or Flanges, multiply the weight factor (lightface subscript) by the weight per foot of covering used on straight pipe.

*All Flanged Fitting, Flanged Valve and Flange weights include the proportional weight of bolts or studs required to make up all joints.

**Cast Iron Valve weights are for flanged valves. Steel Valve weights are for welding end valves.

WEIGHTS OF PIPING MATERIALS—1½" PIPE SIZE

PIPE

Schedule No.	40	80	160	
Wall Designation	Std.	XS		XXS
Thickness—In.	.145	.200	.281	.400
Pipe—Lbs/Ft	2.72	3.63	4.86	6.41
Water—Lbs/Ft	.88	.77	.61	.41

WELDING FITTINGS

	40	80	160	XXS
L.R. 90° Elbow	.8 / .4	1.1 / .4	1.4 / .4	1.8 / .4
S.R. 90° Elbow	.6 / .3	.7 / .3		
L.R. 45° Elbow	.5 / .2	.7 / .2	.8 / .2	1 / .2
Tee	2 / .6	2.5 / .6	3.1 / .6	3.7 / .6
Lateral	4.3 / 1.3	5.8 / 1.3		
Reducer	.6 / .2	.7 / .2	.9 / .2	1.2 / .2
Cap	.4 / .3	.5 / .3	.7 / .3	.7 / .3

COVERING

Temperature Range °F	to 260°	260-360	360-440	440-525	525-600	600-700	700-800	800-900	900-1000
85% Magnesia Thickness—In.	⅛	⅛	1½	2	1¹¹⁄₁₆				
85% Magnesia Lbs/Ft	.85	.85	1.75	2.65	2.45				
Combination Thickness—In.						2	2	2	2
Combination Lbs/Ft						4.2	4.2	4.2	4.2
Calcium Silicate Thickness—In.	1	1	1	1	1½	1½	2	2½	2½
Calcium Silicate Lbs/Ft	.88	.88	.88	.88	1.45	1.45	1.82	2.63	2.63

FLANGES

Pressure Rating psi	Cast Iron		Steel						
	125	250	150	300	400	600	900	1500	2500
Screwed or Slip-On	3.5 / 1.5	7 / 1.5	3.5 / 1.5	8 / 1.5		9 / 1.5		19 / 1.5	31 / 1.5
Welding Neck			4 / 1.5	9 / 1.5		11 / 1.5		17 / 1.5	32 / 1.5
Lap Joint			3.5 / 1.5	8 / 1.5		9 / 1.5		19 / 1.5	31 / 1.5
Blind	3.5 / 1.5	7 / 1.5	3.5 / 1.5	9 / 1.5		10 / 1.5		19 / 1.5	31 / 1.5

FLANGED FITTINGS

	125	250	150	300	400	600	900	1500	2500
S.R. 90° Elbow	10 / 3.7		12 / 3.7	23 / 3.8		26 / 3.9		46 / 4	
L.R. 90° Elbow	12 / 4		13 / 4	24 / 4					
45° Elbow	9 / 3.4		11 / 3.4	21 / 3.5		23 / 3.5		39 / 3.7	
Tee	17 / 5.6		20 / 5.6	30 / 5.7		37 / 5.8		70 / 6	

**VALVES

	125	250	150	300	400	600	900	1500	2500
Flanged Bonnet Gate	27 / 6.8			51 / 4		71 / 4.2		114 / 4.5	
Flanged Bonnet Globe or Angle				40 / 4.1		46 / 4.2		111 / 4.5	
Flanged Bonnet Check				32 / 4.1		33 / 4.2		81 / 4.5	
Pressure Seal Bonnet—Gate							42 / 1.9	42 / 1.2	
Pressure Seal Bonnet—Globe									

Bolts

	125	250	150	300	400	600	900	1500	2500
*One Complete Flanged Joint	1	2.5	1	3.5		3.5		9	12

SEE GENERAL NOTES FOR MATERIALS NOT SHOWN

All weights are shown in bold type.

The weight of steel pipe is per linear foot.

For Boiler Feed Piping, add the weight of water to the weight of steel pipe.

The pipe covering thicknesses and weights indicate the average conditions per linear foot and include all allowances for wire, cement, canvas, bands and paint. The listed thickness of combination covering is the sum of the inner and the outer layer thickness.

Pipe covering temperature ranges are intended as a guide only and do not constitute a recommendation for specific thickness of materials.

To find the weight of covering on Flanged Fittings, Valves, or Flanges, multiply the weight factor (lightface subscript) by the weight per foot of covering used on straight pipe.

*All Flanged Fitting, Flanged Valve and Flange weights include the proportional weight of bolts or studs required to make up all joints.

**Cast Iron Valve weights are for flanged valves. Steel Valve weights are for welding end valves.

WEIGHTS OF PIPING MATERIALS—2″ PIPE SIZE

PIPE

	40	80	160	
Schedule No.	40	80	160	
Wall Designation	Std.	XS		XXS
Thickness—In.	.154	.218	.343	.436
Pipe—Lbs/Ft	3.65	5.02	7.44	9.03
Water—Lbs/Ft	1.46	1.28	.97	.77

WELDING FITTINGS (weight / covering factor)

L.R. 90° Elbow	1.5 / .5	2 / .5	2.9 / .5	3.5 / .5
S.R. 90° Elbow	1 / .3	1.3 / .3		
L.R. 45° Elbow	.8 / .2	1.1 / .2	1.6 / .2	1.8 / .2
Tee	3 / .6	3.7 / .6	5 / .6	5.7 / .6
Lateral	6.6 / 1.4	9.8 / 1.4		
Reducer	.9 / .3	1.2 / .3	1.6 / .3	1.9 / .3
Cap	.5 / .4	.7 / .4	1.2 / .4	1.2 / .4

COVERING

Temperature Range °F	to 260°	260-360	360-440	440-525	525-600	600-700	700-800	800-900	900-1000
85% Magnesia Thickness—In.	1-1/32	1-1/32	1-1/2	2	2-5/32				
85% Magnesia Lbs/Ft	1.25	1.25	2.05	3.15	3.40				
Combination Thickness—In.						2-3/4	2-3/4	3-1/4	3-3/8
Combination Lbs/Ft						5.8	5.8	7.4	9.2
Calcium Silicate Thickness—In.	1	1	1	1-1/2	1-1/2	2	2	2-1/2	2-1/2
Calcium Silicate Lbs/Ft	1.01	1.01	1.01	1.69	1.69	2.50	2.50	3.38	3.38

FLANGES (weight / covering factor)

Pressure Rating psi	Cast Iron 125	Cast Iron 250	Steel 150	Steel 300	Steel 400	Steel 600	Steel 900	Steel 1500	Steel 2500
Screwed or Slip-On	6 / 1.5	/ 1.5	6 / 1.5	/ 1.5		11 / 1.5		32 / 1.5	48 / 1.5
Welding Neck			6 / 1.5	10 / 1.5		13 / 1.5		29 / 1.5	48 / 1.5
Lap Joint			6 / 1.5	9 / 1.5		12 / 1.5		32 / 1.5	48 / 1.5
Blind	6 / 1.5	10 / 1.5	4.8 / 1.5	10 / 1.5		12 / 1.5		31 / 1.5	49 / 1.5

FLANGED FITTINGS (weight / covering factor)

	Cast Iron 125	Cast Iron 250	Steel 150	Steel 300	Steel 400	Steel 600	Steel 900	Steel 1500	Steel 2500
S.R. 90° Elbow	16 / 3.8	24 / 3.8	19 / 3.8	29 / 3.8		35 / 4		83 / 4.2	
L.R. 90° Elbow	18 / 4.1	27 / 4.1	22 / 4.1	31 / 4.1					
45° Elbow	14 / 3.4	22 / 3.5	16 / 3.4	24 / 3.5		33 / 3.7		73 / 3.9	
Tee	23 / 5.7	37 / 5.7	27 / 5.7	41 / 5.7		52 / 6		129 / 6.3	

VALVES (weight / covering factor)

	Cast Iron 125	Cast Iron 250	Steel 150	Steel 300	Steel 400	Steel 600	Steel 900	Steel 1500	Steel 2500
Flanged Bonnet Gate	37 / 6.9	52 / 7.1	43 / 3.9	65 / 4.1		83 / 4.4		154 / 4.8	
Flanged Bonnet Globe or Angle	30 / 7	64 / 7.3	42 / 4	58 / 4.3		78 / 4.4		157 / 4.8	
Flanged Bonnet Check	26 / 7	51 / 7.3	27 / 4	55 / 4.3		47 / 4.4		106 / 4.8	
Pressure Seal Bonnet—Gate							75 / 2.1	75 / 1.4	
Pressure Seal Bonnet—Globe								135 / 2.1	

BOLTS

	Cast Iron 125	Cast Iron 250	Steel 150	Steel 300	Steel 400	Steel 600	Steel 900	Steel 1500	Steel 2500
*One Complete Flanged Joint	1.5	3.5	1.5	4		4.5		12.5	21

SEE GENERAL NOTES FOR MATERIALS NOT SHOWN

All weights are shown in bold type.

The weight of steel pipe is per linear foot.

For Boiler Feed Piping, add the weight of water to the weight of steel pipe.

The pipe covering thicknesses and weights indicate the average conditions per linear foot and include all allowances for wire, cement, canvas, bands and paint. The listed thickness of combination covering is the sum of the inner and the outer layer thickness.

Pipe covering temperature ranges are intended as a guide only and do not constitute a recommendation for specific thickness of materials.

To find the weight of covering on Flanged Fittings, Valves, or Flanges, multiply the weight factor (lightface subscript) by the weight per foot of covering used on straight pipe.

*All Flanged Fitting, Flanged Valve and Flange weights include the proportional weight of bolts or studs required to make up all joints.

**Cast Iron Valve weights are for flanged valves. Steel Valve weights are for welding end valves.

WEIGHTS OF PIPING MATERIALS—2½″ PIPE SIZE

PIPE

Schedule No.	40	80	160	
Wall Designation	Std.	XS		XXS
Thickness—In.	.203	.276	.375	.552
Pipe—Lbs/Ft	5.79	7.66	10.01	13.70
Water—Lbs/Ft	2.08	1.84	1.54	1.07

WELDING FITTINGS

L.R. 90° Elbow	2.9 / .6	3.8 / .6	4.9 / .6	6.5 / .6
S.R. 90° Elbow	1.9 / .4	2.5 / .4		
L.R. 45° Elbow	1.6 / .3	2.1 / .3	2.7 / .3	3.5 / .3
Tee	5.2 / .8	6.4 / .8	7.9 / .8	9.9 / .8
Lateral	11 / 1.5	14.4 / 1.5		
Reducer	1.6 / .3	2.1 / .3	2.7 / .3	3.4 / .3
Cap	.8 / .4	1 / .4	2 / .4	2.1 / .4

COVERING

Temperature Range °F	to 260°	260-360	360-440	440-525	525-600	600-700	700-800	800-900	900-1000	
85% Magnesia Thickness—In.	1½	1½	1½	2	2½					
Lbs/Ft	1.35	1.35	2.30	3.40	3.75					
Combination Thickness—In.						2¹³⁄₁₆	2¹³⁄₁₆	3⁵⁄₁₆	3⁵⁄₁₆	
Lbs/Ft						6.6	6.6	8.5	8.5	
Calcium Silicate Thickness—In.	1	1	1	1½	1½	2	2½	2½	3	
Lbs/Ft	1.15	1.15	1.15	1.53	1.53	2.34	3.22	3.22	4.23	

FLANGES

Pressure Rating psi	Cast Iron		Steel							
	125	250	150	300	400	600	900	1500	2500	
Screwed or Slip-On	8 / 1.5	14 / 1.5	9 / 1.5	14 / 1.5		17 / 1.5		46 / 1.5	68 / 1.5	
Welding Neck			9 / 1.5	14 / 1.5		20 / 1.5		42 / 1.5	59 / 1.5	
Lap Joint			9 / 1.5	14 / 1.5		18 / 1.5		46 / 1.5	67 / 1.5	
Blind	8 / 1.5	15 / 1.5	9 / 1.5	16 / 1.5		19 / 1.5		45 / 1.5	69 / 1.5	

FLANGED FITTINGS

	125	250	150	300	400	600	900	1500	2500	
S.R. 90° Elbow	21 / 3.8	36 / 3.9	27 / 3.8	42 / 3.9		50 / 4.1		114 / 4.4		
L.R. 90° Elbow	25 / 4.2	40 / 4.2	30 / 4.2	47 / 4.2						
45° Elbow	19 / 3.5	34 / 3.6	22 / 3.5	35 / 3.6		46 / 3.8		99 / 3.9		
Tee	32 / 5.7	55 / 5.8	42 / 5.7	61 / 5.9		77 / 6.2		169 / 6.6		

VALVES**

	125	250	150	300	400	600	900	1500	2500	
Flanged Bonnet Gate	50 / 7	82 / 7.1	53 / 4	83 / 4.1		108 / 4.6		221 / 5.1		
Flanged Bonnet Globe or Angle	43 / 7.1	87 / 7.4	50 / 4.1	84 / 4.4		98 / 4.6		242 / 5.1		
Flanged Bonnet Check	36 / 7.1	71 / 7.4	32 / 4.1	68 / 4.4		68 / 4.6		175 / 5.1		
Pressure Seal Bonnet—Gate							100 / 2.3	100 / 1.7		
Pressure Seal Bonnet—Globe								180 / 2.3		

Bolts

*One Complete Flanged Joint	1.5	6	1.5	7		8		19	27	

SEE GENERAL NOTES FOR MATERIALS NOT SHOWN

All weights are shown in bold type.

The weight of steel pipe is per linear foot.

For Boiler Feed Piping, add the weight of water to the weight of steel pipe.

The pipe covering thicknesses and weights indicate the average conditions per linear foot and include all allowances for wire, cement, canvas, bands and paint. The listed thickness of combination covering is the sum of the inner and the outer layer thickness.

Pipe covering temperature ranges are intended as a guide only and do not constitute a recommendation for specific thickness of materials.

To find the weight of covering on Flanged Fittings, Valves, or Flanges, multiply the weight factor (lightface subscript) by the weight per foot of covering used on straight pipe.

*All Flanged Fitting, Flanged Valve and Flange weights include the proportional weight of bolts or studs required to make up all joints.

**Cast Iron Valve weights are for flanged valves. Steel Valve weights are for welding end valves.

WEIGHTS OF PIPING MATERIALS—3" PIPE SIZE

PIPE

Schedule No.	40	80	160	
Wall Designation	Std.	XS		XXS
Thickness—In.	.216	.300	.437	.600
Pipe—Lbs/Ft	7.58	10.25	14.32	18.58
Water—Lbs/Ft	3.20	2.86	2.35	1.80

WELDING FITTINGS

Fitting				
L.R. 90° Elbow	4.6 / .8	6.1 / .8	8.4 / .8	10.7 / .8
S.R. 90° Elbow	3 / .5	4 / .5		
L.R. 45° Elbow	2.4 / .3	3.2 / .3	4.4 / .3	5.4 / .3
Tee	7.4 / .8	9.5 / .8	12.2 / .8	14.8 / .8
Lateral	17 / 1.8	24 / 1.8		
Reducer	2.2 / .3	2.9 / .3	3.7 / .3	4.7 / .3
Cap	1.4 / .5	1.8 / .5	3.5 / .5	3.7 / .5

COVERING

Temperature Range °F	to 260°	260-360	360-440	440-525	525-600	600-700	700-800	800-900	900-1000
85% Magnesia Thickness—In.	1½	1½	1½	2	2½				
85% Magnesia Lbs/Ft	1.55	1.55	2.65	3.95	4.25				
Combination Thickness—In.						3⅛	3⅛	3⅜	3⅜
Combination Lbs/Ft						8.3	8.3	10.7	10.7
Calcium Silicate Thickness—In.	1	1	1	1½	1½	2	2½	2½	3
Calcium Silicate Lbs/Ft	1.28	1.28	1.28	2.09	2.09	2.98	3.98	3.98	5.11

FLANGES

Pressure Rating psi	Cast Iron		Steel						
	125	250	150	300	400	600	900	1500	2500
Screwed or Slip-On	9 / 1.5	17 / 1.5	10 / 1.5	17 / 1.5		20 / 1.5	38 / 1.5	61 / 1.5	101 / 1.5
Welding Neck			10 / 1.5	17 / 1.5		24 / 1.5	35 / 1.5	54 / 1.5	101 / 1.5
Lap Joint			10 / 1.5	17 / 1.5		21 / 1.5	38 / 1.5	62 / 1.5	100 / 1.5
Blind	10 / 1.5	19 / 1.5	10 / 1.5	17 / 1.5		24 / 1.5	39 / 1.5	61 / 1.5	104 / 1.5

FLANGED FITTINGS

	125	250	150	300	400	600	900	1500	2500
S.R. 90° Elbow	26 / 3.9	46 / 4	32 / 3.9	53 / 4		67 / 4.1	98 / 4.3	150 / 4.6	
L.R. 90° Elbow	30 / 4.3	50 / 4.3	40 / 4.3	63 / 4.3					
45° Elbow	22 / 3.5	41 / 3.6	28 / 3.5	46 / 3.6		60 / 3.8	93 / 3.9	135 / 4	
Tee	39 / 5.9	67 / 6	52 / 5.9	81 / 6		102 / 6.2	151 / 6.5	238 / 6.9	

VALVES

	125	250	150	300	400	600	900	1500	2500
Flanged Bonnet Gate	66 / 7	112 / 7.4	77 / 4	119 / 4.4		153 / 4.8	225 / 4.9	338 / 5.3	
Flanged Bonnet Globe or Angle	56 / 7.2	121 / 7.6	80 / 4.2	102 / 4.6		132 / 4.8	242 / 4.9	341 / 5.3	
Flanged Bonnet Check	46 / 7.2	100 / 7.6	51 / 4.2	101 / 4.6		91 / 4.8	146 / 4.9	233 / 5.3	
Pressure Seal Bonnet—Gate							140 / 2.5	140 / 2.5	
Pressure Seal Bonnet—Globe							160 / 2.5	260 / 2.5	

Bolts

	125	250	150	300	400	600	900	1500	2500
*One Complete Flanged Joint	1.5	6	1.5	7.5		8	12.5	25	37

SEE GENERAL NOTES FOR MATERIALS NOT SHOWN

All weights are shown in bold type.

The weight of steel pipe is per linear foot.

For Boiler Feed Piping, add the weight of water to the weight of steel pipe.

The pipe covering thicknesses and weights indicate the average conditions per linear foot and include all allowances for wire, cement, canvas, bands and paint. The listed thickness of combination covering is the sum of the inner and the outer layer thickness.

Pipe covering temperature ranges are intended as a guide only and do not constitute a recommendation for specific thickness of materials.

To find the weight of covering on Flanged Fittings, Valves, or Flanges, multiply the weight factor (lightface subscript) by the weight per foot of covering used on straight pipe.

*All Flanged Fitting, Flanged Valve and Flange weights include the proportional weight of bolts or studs required to make up all joints.

**Cast Iron Valve weights are for flanged valves. Steel Valve weights are for welding end valves.

WEIGHTS OF PIPING MATERIALS—3½″ PIPE SIZE

PIPE	Schedule No.	40	80											
	Wall Designation	Std.	XS	XXS										
	Thickness—In.	.226	.318	.636										
	Pipe—Lbs/Ft	9.11	12.51	22.85										
	Water—Lbs/Ft	4.28	3.85	2.53										

WELDING FITTINGS

	Schedule 40	80	XXS
L.R. 90° Elbow	6.4 / .9	8.7 / .9	15.4 / .9
S.R. 90° Elbow	4.3 / .6	5.8 / .6	
L.R. 45° Elbow	3.3 / .4	4.4 / .4	7.5 / .4
Tee	9.9 / .9	12.6 / .9	20 / .9
Lateral	22 / 1.8		
Reducer	3.1 / .3	4.1 / .3	6.9 / .3
Cap	2.1 / .6	2.8 / .6	5.5 / .6

COVERING

	Temperature Range °F	to 260°	260-360	360-440	440-525	525-600	600-700	700-800	800-900	900-1000
85% Magnesia	Thickness—In.	1½	1½	1½	2	2¼				
	Lbs/Ft	1.70	1.70	2.90	4.25	4.65				
Combination	Thickness—In.						2¹³⁄₁₆	2¹³⁄₁₆	3⁵⁄₁₆	3⁵⁄₁₆
	Lbs/Ft						7.8	7.8	10.2	10.2
Calcium Silicate	Thickness—In.	1	1	1½	1½	2	2	2½	3	3
	Lbs/Ft	1.06	1.06	1.86	1.86	2.75	2.75	3.75	4.88	4.88

FLANGES / FLANGED FITTINGS / VALVES / Bolts

		Cast Iron		Steel						
Pressure Rating psi		125	250	150	300	400	600	900	1500	2500
Screwed or Slip-On		13 / 1.5	21 / 1.5	13 / 1.5			27 / 1.5			
Welding Neck				12 / 1.5			28 / 1.5			
Lap Joint				14 / 1.5			28 / 1.5			
Blind		14 / 1.5	23 / 1.5	15 / 1.5			35 / 1.5			
S.R. 90° Elbow		35 / 4	56 / 4.1	49 / 4			82 / 4.3			
L.R. 90° Elbow		40 / 4.4	62 / 4.4	54 / 4.4						
45° Elbow		31 / 3.6	51 / 3.7	39 / 3.6			75 / 3.9			
Tee		54 / 6	86 / 6.2	70 / 6			133 / 6.4			
Flanged Bonnet Gate		82 / 7.1	143 / 7.5	88 / 4.1			201 / 4.9			
Flanged Bonnet Globe or Angle		74 / 7.3	137 / 7.7	99 / 4.3			160 / 4.9			
Flanged Bonnet Check		71 / 7.3	125 / 7.7	54 / 4.3			123 / 4.9			
Pressure Seal Bonnet—Gate										
Pressure Seal Bonnet—Globe										
*One Complete Flange Joint		3.5	6.5	3.5			12			

SEE GENERAL NOTES FOR MATERIALS NOT SHOWN

All weights are shown in bold type.

The weight of steel pipe is per linear foot.

For Boiler Feed Piping, add the weight of water to the weight of steel pipe.

The pipe covering thicknesses and weights indicate the average conditions per linear foot and include all allowances for wire, cement, canvas, bands and paint. The listed thickness of combination covering is the sum of the inner and the outer layer thickness.

Pipe covering temperature ranges are intended as a guide only and do not constitute a recommendation for specific thickness of materials.

To find the weight of covering on Flanged Fittings, Valves, or Flanges, multiply the weight factor (lightface subscript) by the weight per foot of covering used on straight pipe.

*All Flanged Fitting, Flanged Valve and Flange weights include the proportional weight of bolts or studs required to make up all joints.

**Cast Iron Valve weights are for flanged valves. Steel Valve weights are for welding end valves.

WEIGHTS OF PIPING MATERIALS—4″ PIPE SIZE

PIPE

Schedule No.	40	80	120	160	
Wall Designation	Std.	XS			XXS
Thickness—In.	.237	.337	.437	.531	.674
Pipe—Lbs/Ft	10.79	14.98	18.96	22.51	27.54
Water—Lbs/Ft	5.51	4.98	4.48	4.02	3.38

WELDING FITTINGS

	40	80	120	160	XXS
L.R. 90° Elbow	8.7 / 1	11.9 / 1		17.6 / 1	21 / 1
S.R. 90° Elbow	5.8 / .7	7.9 / .7			
L.R. 45° Elbow	4.3 / .4	5.9 / .4		8.5 / .4	10.1 / .4
Tee	12.6 / 1	16.4 / 1		23 / 1	27 / 1
Lateral	30 / 2.1	45 / 2.1			
Reducer	3.6 / .3	4.9 / .3		6.6 / .3	8.2 / .3
Cap	2.6 / .6	3.4 / .6		6.5 / .6	6.7 / .6

COVERING

Temperature Range °F	to 260°	260-360	360-440	440-525	525-600	600-700	700-800	800-900	900-1000
85% Magnesia — Thickness—In.	1⅛	1⅛	1½	2	2¼				
85% Magnesia — Lbs/Ft	2.10	2.10	3.15	4.75	5.10				
Combination — Thickness—In.						3 1/16	3 1/16	3 3/16	3 3/16
Combination — Lbs/Ft						9.8	9.8	12.2	12.2
Calcium Silicate — Thickness—In.	1	1	1¼	1½	2	2	2½	3	3
Calcium Silicate — Lbs/Ft	1.60	1.60	2.49	2.49	3.49	3.49	4.62	6.03	6.03

FLANGES / FLANGED FITTINGS / VALVES / BOLTS

Pressure Rating psi	Cast Iron		Steel						
	125	250	150	300	400	600	900	1500	2500
Screwed or Slip-On	16 / 1.5	26 / 1.5	15 / 1.5	26 / 1.5	32 / 1.5	43 / 1.5	66 / 1.5	94 / 1.5	158 / 1.5
Welding Neck			14 / 1.5	26 / 1.5	37 / 1.5	43 / 1.5	57 / 1.5	81 / 1.5	159 / 1.5
Lap Joint			16 / 1.5	27 / 1.5	33 / 1.5	45 / 1.5	67 / 1.5	94 / 1.5	155 / 1.5
Blind	18 / 1.5	29 / 1.5	19 / 1.5	31 / 1.5	39 / 1.5	47 / 1.5	66 / 1.5	90 / 1.5	164 / 1.5
S.R. 90° Elbow	45 / 4.1	72 / 4.2	59 / 4.1	85 / 4.2	99 / 4.3	128 / 4.4	185 / 4.5	254 / 4.8	
L.R. 90° Elbow	52 / 4.5	79 / 4.5	72 / 4.5	98 / 4.5					
45° Elbow	40 / 3.7	65 / 3.8	51 / 3.7	78 / 3.8	82 / 3.9	119 / 4	170 / 4.1	214 / 4.2	
Tee	70 / 6.1	109 / 6.3	86 / 6.1	121 / 6.3	153 / 6.4	187 / 6.6	262 / 6.8	386 / 7.2	
Flanged Bonnet Gate	109 / 7.2	188 / 7.5	114 / 4.2	173 / 4.5	213 / 5	274 / 5.1	370 / 5.3	566 / 5.7	
Flanged Bonnet Globe or Angle	97 / 7.4	177 / 7.8	127 / 4.4	168 / 4.8	194 / 5	222 / 5.1	383 / 5.3	546 / 5.7	
Flanged Bonnet Check	80 / 7.4	146 / 7.8	104 / 4.4	146 / 4.8	180 / 5	159 / 5.1	256 / 5.3	344 / 5.7	
Pressure Seal Bonnet—Gate							230 / 2.8	235 / 3	
Pressure Seal Bonnet—Globe							260 / 2.8	375 / 3	
*One Complete Flanged Joint	4	6.5	4	7.5	12	12.5	25	34	61

SEE GENERAL NOTES FOR MATERIALS NOT SHOWN

All weights are shown in bold type.

The weight of steel pipe is per linear foot.

For Boiler Feed Piping, add the weight of water to the weight of steel pipe.

The pipe covering thicknesses and weights indicate the average conditions per linear foot and include all allowances for wire, cement, canvas, bands and paint. The listed thickness of combination covering is the sum of the inner and the outer layer thickness.

Pipe covering temperature ranges are intended as a guide only and do not constitute a recommendation for specific thickness of materials.

To find the weight of covering on Flanged Fittings, Valves, or Flanges, multiply the weight factor (lightface subscript) by the weight per foot of covering used on straight pipe.

*All Flanged Fitting, Flanged Valve and Flange weights include the proportional weight of bolts or studs required to make up all joints.

**Cast Iron Valve weights are for flanged valves. Steel Valve weights are for welding end valves.

WEIGHTS OF PIPING MATERIALS—5″ PIPE SIZE

PIPE

Schedule No.	40	80	120	160						
Wall Designation	Std.	XS			XXS					
Thickness—In.	.258	.375	.500	.625	.750					
Pipe—Lbs/Ft	14.62	20.78	27.04	32.96	38.55					
Water—Lbs/Ft	8.66	7.89	7.09	6.33	5.62					

WELDING FITTINGS

L.R. 90° Elbow	14.7 / 1.3	21 / 1.3			32 / 1.3	37 / 1.3				
S.R. 90° Elbow	9.8 / .8	13.7 / .8								
L.R. 45° Elbow	7.3 / .5	10.2 / .5			15.6 / .5	17.7 / .5				
Tee	19.8 / 1.2	26 / 1.2			39 / 1.2	43 / 1.2				
Lateral	49 / 2.5	70 / 2.5								
Reducer	6 / .4	8.3 / .4			12.4 / .4	14.2 / .4				
Cap	4.2 / .7	5.7 / .7			11 / .7	11 / .7				

COVERING

Temperature Range °F	to 260°	260-360	360-440	440-525	525-600	600-700	700-800	800-900	900-1000	
85% Magnesia — Thickness—In.	1⅛	1½	2	2½	3					
85% Magnesia — Lbs/Ft	2.50	3.75	5.60	7.40	9.30					
Combination — Thickness—In.						3	3½	4	4½	5 5½ 6
Combination — Lbs/Ft						10.9	13.3	16.1	20.6	25.6 29.7 34.1
Calcium Silicate — Thickness—In.	1	1	1½	1½	2	2	2½	3	3	3½ 4 4½
Calcium Silicate — Lbs/Ft	1.84	1.84	2.84	2.84	3.97	3.97	5.37	6.75	6.75	8.26 10.3 12.1

FLANGES / FLANGED FITTINGS / VALVES / BOLTS

Pressure Rating psi	Cast Iron		Steel							
	125	250	150	300	400	600	900	1500	2500	
Screwed or Slip-On	20 / 1.5	32 / 1.5	18 / 1.5	32 / 1.5	37 / 1.5	73 / 1.5	100 / 1.5	172 / 1.5	259 / 1.5	
Welding Neck			18 / 1.5	31 / 1.5	42 / 1.5	70 / 1.5	94 / 1.5	145 / 1.5	263 / 1.5	
Lap Joint			19 / 1.5	33 / 1.5	39 / 1.5	75 / 1.5	101 / 1.5	171 / 1.5	257 / 1.5	
Blind	23 / 1.5	37 / 1.5	23 / 1.5	39 / 1.5	50 / 1.5	78 / 1.5	104 / 1.5	172 / 1.5	272 / 1.5	
S.R. 90° Elbow	58 / 4.3	94 / 4.3	80 / 4.3	113 / 4.3	123 / 4.5	205 / 4.7	268 / 4.8	435 / 5.2		
L.R. 90° Elbow	68 / 4.7	105 / 4.7	91 / 4.7	128 / 4.7						
45° Elbow	51 / 3.8	83 / 3.8	66 / 3.8	98 / 3.8	123 / 4	180 / 4.2	239 / 4.3	350 / 4.5		
Tee	90 / 6.4	145 / 6.5	119 / 6.4	172 / 6.4	179 / 6.8	304 / 7	415 / 7.2	665 / 7.8		
Flanged Bonnet Gate	138 / 7.3	264 / 7.9	151 / 4.3	257 / 4.9	309 / 5.3	386 / 5.5	508 / 5.8	841 / 6.3		
Flanged Bonnet Globe or Angle	138 / 7.6	247 / 8	172 / 4.6	237 / 5	277 / 5.3	274 / 5.5	658 / 5.8			
Flanged Bonnet Check	118 / 7.6	210 / 8	141 / 4.6	198 / 5	249 / 5.3	244 / 5.5	326 / 5.8	531 / 6.3		
Pressure Seal Bonnet—Gate							350 / 3.1	370 / 3.4		
Pressure Seal Bonnet—Globe							395 / 3.1	500 / 3.4		
*One Complete Flanged Joint	6	6.5	6	8	12.5	19.5	33	60	98	

SEE GENERAL NOTES FOR MATERIALS NOT SHOWN

All weights are shown in bold type.

The weight of steel pipe is per linear foot.

For Boiler Feed Piping, add the weight of water to the weight of steel pipe.

The pipe covering thicknesses and weights indicate the average conditions per linear foot and include all allowances for wire, cement, canvas, bands and paint. The listed thickness of combination covering is the sum of the inner and the outer layer thickness.

Pipe covering temperature ranges are intended as a guide only and do not constitute a recommendation for specific thickness of materials.

To find the weight of covering on Flanged Fittings, Valves, or Flanges, multiply the weight factor (lightface subscript) by the weight per foot of covering used on straight pipe.

*All Flanged Fitting, Flanged Valve and Flange weights include the proportional weight of bolts or studs required to make up all joints.

**Cast Iron Valve weights are for flanged valves. Steel Valve weights are for welding end valves.

WEIGHTS OF PIPING MATERIALS—6" PIPE SIZE

PIPE

Schedule No.	40	80	120	160	
Wall Designation	Std.	XS			XXS
Thickness—In.	.280	.432	.562	.718	.864
Pipe—Lbs/Ft	18.97	28.57	36.39	45.3	53.2
Water—Lbs/Ft	12.51	11.29	10.30	9.2	8.2

WELDING FITTINGS

(weight in bold / covering factor in lightface)

L.R. 90° Elbow	23 / 1.5	34 / 1.5		53 / 1.5	62 / 1.5
S.R. 90° Elbow	15.2 / 1	23 / 1			
L.R. 45° Elbow	11.3 / .6	16.7 / .6		26 / .6	30 / .6
Tee	29.3 / 1.4	42 / 1.4		60 / 1.4	68 / 1.4
Lateral	79 / 2.9	101 / 2.9			
Reducer	8.7 / .5	12.6 / .5		18.8 / .5	21 / .5
Cap	6.4 / .9	9.2 / .9		17.5 / .9	17.5 / .9

COVERING

Temperature Range °F	to 260°	260-360	360-440	440-525	525-600	600-700	700-800	800-900	900-1000			
85% Magnesia Thickness—In.	1½	1½	2	2½	3							
85% Magnesia Lbs/Ft	2.90	4.15	6.40	8.40	10.0							
Combination Thickness—In.						3	3½	4	4½	5	5½	6
Combination Lbs/Ft						12.3	14.9	18.2	24.2	28.2	32.6	37.4
Calcium Silicate Thickness—In.	1½	1½	1½	1½	2	2	2½	3	3	3½	4	4½
Calcium Silicate Lbs/Ft	3.13	3.13	3.13	3.13	4.54	4.54	5.92	7.42	7.42	9.47	11.2	13.1

FLANGES, FLANGED FITTINGS, VALVES, BOLTS

(weight in bold / covering factor in lightface)

Pressure Rating psi	Cast Iron 125	Cast Iron 250	Steel 150	Steel 300	Steel 400	Steel 600	Steel 900	Steel 1500	Steel 2500
FLANGES									
Screwed or Slip-On	25 / 1.5	42 / 1.5	22 / 1.5	45 / 1.5	54 / 1.5	95 / 1.5	128 / 1.5	202 / 1.5	395 / 1.5
Welding Neck			22 / 1.5	42 / 1.5	56 / 1.5	85 / 1.5	116 / 1.5	176 / 1.5	402 / 1.5
Lap Joint			24 / 1.5	47 / 1.5	56 / 1.5	98 / 1.5	131 / 1.5	213 / 1.5	393 / 1.5
Blind	28 / 1.5	51 / 1.5	29 / 1.5	56 / 1.5	71 / 1.5	101 / 1.5	133 / 1.5	197 / 1.5	417 / 1.5
FLANGED FITTINGS									
S.R. 90° Elbow	74 / 4.3	125 / 4.4	90 / 4.3	147 / 4.4	184 / 4.6	275 / 4.8	375 / 5	566 / 5.3	
L.R. 90° Elbow	91 / 4.9	145 / 4.9	126 / 4.9	182 / 4.9					
45° Elbow	66 / 3.8	115 / 3.9	82 / 3.8	132 / 3.9	149 / 4.1	240 / 4.3	320 / 4.3	476 / 4.6	
Tee	114 / 6.5	195 / 6.6	149 / 6.5	217 / 6.6	279 / 6.9	400 / 7.2	565 / 7.5	839 / 8	
VALVES									
Flanged Bonnet Gate	172 / 7.3	359 / 8	210 / 4.3	367 / 5	409 / 5.5	553 / 5.8	784 / 6	1227 / 6.6	
Flanged Bonnet Globe or Angle	184 / 7.8	345 / 8.2	238 / 4.8	333 / 5.2	366 / 5.4	465 / 5.8	844 / 6		
Flanged Bonnet Check	154 / 7.8	286 / 8.2	176 / 4.8	272 / 5.2	341 / 5.4	335 / 5.8	459 / 6	877 / 6.5	
Pressure Seal Bonnet—Gate							540 / 3.5	600 / 3.8	
Pressure Seal Bonnet—Globe							600 / 3.5	700 / 3.8	
Bolts *One Complete Flanged Joint	6	10	6	11.5	19	30	40	76	145

SEE GENERAL NOTES FOR MATERIALS NOT SHOWN

All weights are shown in bold type.

The weight of steel pipe is per linear foot.

For Boiler Feed Piping, add the weight of water to the weight of steel pipe.

The pipe covering thicknesses and weights indicate the average conditions per linear foot and include all allowances for wire, cement, canvas, bands and paint. The listed thickness of combination covering is the sum of the inner and the outer layer thickness.

Pipe covering temperature ranges are intended as a guide only and do not constitute a recommendation for specific thickness of materials.

To find the weight of covering on Flanged Fittings, Valves, or Flanges, multiply the weight factor (lightface subscript) by the weight per foot of covering used on straight pipe.

*All Flanged Fitting, Flanged Valve and Flange weights include the proportional weight of bolts or studs required to make up all joints.

**Cast Iron Valve weights are for flanged valves. Steel Valve weights are for welding end valves.

WEIGHTS OF PIPING MATERIALS—8" PIPE SIZE

PIPE

	20	30	40	60	80	100	120	140		160
Schedule No.	20	30	40	60	80	100	120	140		160
Wall Designation			Std.		XS				XXS	
Thickness—In.	.250	.277	.322	.406	.500	.593	.718	.812	.875	.906
Pipe—Lbs/Ft	22.36	24.70	28.55	35.64	43.4	50.9	60.6	67.8	72.4	74.7
Water—Lbs/Ft	22.48	22.18	21.69	20.79	19.8	18.8	17.6	16.7	16.1	15.8

WELDING FITTINGS

(weight / weight factor)

	40	80	(XXS)	160
L.R. 90° Elbow	46 / 2	69 / 2	114 / 2	117 / 2
S.R. 90° Elbow	31 / 1.3	46 / 1.3		
L.R. 45° Elbow	23 / .8	34 / .8	55 / .8	56 / .8
Tee	54 / 1.8	76 / 1.8	118 / 1.8	120 / 1.8
Lateral	155 / 3.8	216 / 3.8		
Reducer	13.9 / .5	20 / .5	32 / .5	33 / .5
Cap	11.3 / 1	16.3 / 1	31 / 1	32 / 1

COVERING

Temperature Range °F	to 260°	260-360	360-440	440-525	525-600	600-700	700-800	800-900	900-1000			
85% Magnesia — Thickness—In.	1¼	1½	2	2½	3							
85% Magnesia — Lbs/Ft	4.05	5.30	7.70	10.3	12.5							
Combination — Thickness—In.						3	3½	4	4½	5	5½	6
Combination — Lbs/Ft						16.9	19.6	23.1	26.7	32.8	38.3	43.4
Calcium Silicate — Thickness—In.	1½	1½	1½	1½	2	2	2½	3	3½	4	4½	5
Calcium Silicate — Lbs/Ft	4.06	4.06	4.06	4.06	5.56	5.56	7.61	9.38	11.3	13.3	15.4	17.6

FLANGES, FLANGED FITTINGS, VALVES and BOLTS

(weight / weight factor)

Pressure Rating psi	Cast Iron 125	Cast Iron 250	Steel 150	Steel 300	Steel 400	Steel 600	Steel 900	Steel 1500	Steel 2500
FLANGES									
Screwed or Slip-On	34 / 1.5	64 / 1.5	33 / 1.5	67 / 1.5	82 / 1.5	135 / 1.5	206 / 1.5	319 / 1.5	601 / 1.5
Welding Neck			33 / 1.5	66 / 1.5	87 / 1.5	117 / 1.5	193 / 1.5	280 / 1.6	613 / 1.5
Lap Joint			36 / 1.5	70 / 1.5	86 / 1.5	139 / 1.5	227 / 1.5	354 / 1.5	595 / 1.5
Blind	45 / 1.5	83 / 1.5	48 / 1.5	90 / 1.5	115 / 1.5	159 / 1.5	231 / 1.5	362 / 1.5	649 / 1.5
FLANGED FITTINGS									
S.R. 90° Elbow	117 / 4.5	201 / 4.7	157 / 4.5	238 / 4.7	310 / 5	435 / 5.2	639 / 5.4	995 / 5.7	
L.R. 90° Elbow	152 / 5.3	236 / 5.3	202 / 5.3	283 / 5.3					
45° Elbow	101 / 3.9	171 / 4	127 / 3.9	203 / 4	215 / 4.1	360 / 4.4	507 / 4.5	870 / 4.8	
Tee	175 / 6.8	304 / 7.1	230 / 6.8	337 / 7.1	445 / 7.5	610 / 7.8	978 / 8.1	1465 / 8.6	
VALVES									
Flanged Bonnet Gate	251 / 7.5	583 / 8.1	329 / 4.5	549 / 5.1	727 / 6	1008 / 6.3	1332 / 6.6		
Flanged Bonnet Globe or Angle	317 / 8.4	554 / 8.6	408 / 5.4	509 / 5.6	576 / 5.9	1200 / 6.3			
Flanged Bonnet Check	302 / 8.4	454 / 8.6	301 / 5.4	467 / 5.6	561 / 5.9	563 / 6.3	677 / 6.6		
Pressure Seal Bonnet—Gate							835 / 4.3	975 / 4.5	
Pressure Seal Bonnet—Globe							1000 / 4.3	1115 / 4.5	
BOLTS									
*One Complete Flanged Joint	6.5	16	6.5	18	30	40	69	121	232

SEE GENERAL NOTES FOR MATERIALS NOT SHOWN

All weights are shown in bold type.

The weight of steel pipe is per linear foot.

For Boiler Feed Piping, add the weight of water to the weight of steel pipe.

The pipe covering thicknesses and weights indicate the average conditions per linear foot and include all allowances for wire, cement, canvas, bands and paint. The listed thickness of combination covering is the sum of the inner and the outer layer thickness.

Pipe covering temperature ranges are intended as a guide only and do not constitute a recommendation for specific thickness of materials.

To find the weight of covering on Flanged Fittings, Valves, or Flanges, multiply the weight factor (lightface subscript) by the weight per foot of covering used on straight pipe.

*All Flanged Fitting, Flanged Valve and Flange weights include the proportional weight of bolts or studs required to make up all joints.

**Cast Iron Valve weights are for flanged valves. Steel Valve weights are for welding end valves.

WEIGHTS OF PIPING MATERIALS—10" PIPE SIZE

PIPE

Schedule No.	20	30	40	60	80	100	120	140	160
Wall Designation			Std.		XS				
Thickness—In	.250	.307	.365	.500	.593	.718	.843	1.000	1.125
Pipe—Lbs/Ft	28.04	34.24	40.5	54.7	64.3	76.9	89.2	104.1	115.7
Water—Lbs/Ft	35.77	34.98	34.1	32.3	31.1	29.5	28.0	26.1	24.6

WELDING FITTINGS

	40	80	160
L.R. 90° Elbow	82 / 2.5	109 / 2.5	226 / 2.5
S.R. 90° Elbow	54 / 1.7	73 / 1.7	
L.R. 45° Elbow	40 / 1	54 / 1	109 / 1
Tee	91 / 2.1	118 / 2.1	222 / 2.1
Lateral	238 / 4.4	335 / 4.4	
Reducer	23 / .6	31 / .6	58 / .6
Cap	20 / 1.3	26 / 1.3	54 / 1.3

COVERING

Temperature Range °F	to 260°	260-360	360-440	440-525	525-600	600-700	700-800	800-900	900-1000			
85% Magnesia Thickness—In	1¼	1½	2	2½	3							
85% Magnesia Lbs/Ft	5.20	6.60	9.50	12.3	14.2							
Combination Thickness—In					3	3 9/16	4 3/16	4½	5	5½	6	
Combination Lbs/Ft					20.5	23.5	28.0	33.3	38.5	43.9	49.8	
Calcium Silicate Thickness—In	1½	1½	1½	1½	2	2	2½	3	3½	4	4½	5
Calcium Silicate Lbs/Ft	5.10	5.10	5.10	5.10	6.87	6.87	8.76	10.8	12.9	15.2	17.5	20.0

FLANGES / FLANGED FITTINGS / VALVES / Bolts

Pressure Rating psi	Cast Iron 125	Cast Iron 250	Steel 150	Steel 300	Steel 400	Steel 600	Steel 900	Steel 1500	Steel 2500
Screwed or Slip-On	53 / 1.5	97 / 1.5	51 / 1.5	100 / 1.5	117 / 1.5	213 / 1.5	292 / 1.5	528 / 1.5	1148 / 1.5
Welding Neck			46 / 1.5	95 / 1.5	128 / 1.5	196 / 1.5	267 / 1.5	447 / 1.5	1129 / 1.5
Lap Joint			54 / 1.5	114 / 1.5	143 / 1.5	236 / 1.5	332 / 1.5	589 / 1.5	1131 / 1.5
Blind	71 / 1.5	136 / 1.5	78 / 1.5	146 / 1.5	180 / 1.5	267 / 1.5	337 / 1.5	599 / 1.5	1248 / 1.5
S.R. 90° Elbow	190 / 4.8	323 / 4.9	240 / 4.8	343 / 4.9	462 / 5.2	747 / 5.6	995 / 5.8		
L.R. 90° Elbow	245 / 5.8	383 / 5.8	290 / 5.8	438 / 5.8					
45° Elbow	160 / 4.1	273 / 4.2	185 / 4.1	288 / 4.2	332 / 4.3	572 / 4.6	732 / 4.7		
Tee	293 / 7.2	479 / 7.4	353 / 7.2	527 / 7.4	578 / 7.8	1007 / 8.4	1417 / 8.7		
Flanged Bonnet Gate	471 / 7.7	899 / 8.3	513 / 4.7	888 / 5.3	1193 / 6.3	1571 / 6.9	2511 / 7.1		
Flanged Bonnet Globe or Angle	541 / 9.1	943 / 9.1		993 / 6.1	1068 / 6.8	1346 / 6.9	2586 / 7.1		
Flanged Bonnet Check	453 / 9.1	751 / 9.1	413 / 6	586 / 6.1	718 / 6.3	746 / 6.9			
Pressure Seal Bonnet—Gate							1400 / 4.9	1650 / 5.2	
Pressure Seal Bonnet—Globe							1800 / 4.9	1910 / 5.2	
*One Complete Flanged Joint	15	33	15	38	52	72	95	184	445

SEE GENERAL NOTES FOR MATERIALS NOT SHOWN

All weights are shown in bold type.

The weight of steel pipe is per linear foot.

For Boiler Feed Piping, add the weight of water to the weight of steel pipe.

The pipe covering thicknesses and weights indicate the average conditions per linear foot and include all allowances for wire, cement, canvas, bands and paint. The listed thickness of combination covering is the sum of the inner and the outer layer thickness.

Pipe covering temperature ranges are intended as a guide only and do not constitute a recommendation for specific thickness of materials.

To find the weight of covering on Flanged Fittings, Valves, or Flanges, multiply the weight factor (lightface subscript) by the weight per foot of covering used on straight pipe.

*All Flanged Fitting, Flanged Valve and Flange weights include the proportional weight of bolts or studs required to make up all joints.

**Cast Iron Valve weights are for flanged valves. Steel Valve weights are for welding end valves.

WEIGHTS OF PIPING MATERIALS—12″ PIPE SIZE

PIPE

	20	30		40		60	80	100	120	140	160
Schedule No.	20	30		40		60	80	100	120	140	160
Wall Designation			Std.		XS						
Thickness—In.	.250	.330	.375	.406	.500	.562	.687	.843	1.000	1.125	1.312
Pipe—Lbs/Ft	33.38	43.8	49.6	53.5	65.4	73.2	88.5	107.2	125.5	139.7	160.3
Water—Lbs/Ft	51.10	49.7	49.0	48.5	47.0	46.0	44.0	41.6	39.3	37.5	34.9

WELDING FITTINGS

	.375 (Std)	.500 (XS)	1.312 (160)
L.R. 90° Elbow	119 / 3	157 / 3	375 / 3
S.R. 90° Elbow	80 / 2	104 / 2	
L.R. 45° Elbow	60 / 1.3	78 / 1.3	181 / 1.3
Tee	132 / 2.5	167 / 2.5	360 / 2.5
Lateral	337 / 5.4	556 / 5.4	
Reducer	33 / .7	44 / .7	94 / .7
Cap	30 / 1.5	38 / 1.5	89 / 1.5

COVERING

Temperature Range °F	to 260°	260-360	360-440	440-525	525-600	600-700	700-800	800-900	900-1000			
85% Magnesia — Thickness—In.	1½	2	2½	3	3							
85% Magnesia — Lbs/Ft	9.60	12.5	15.5	18.8	18.8							
Combination — Thickness—In.						$3\frac{3}{16}$	4	$4\frac{1}{8}$	$4\frac{1}{2}$	5	$5\frac{1}{2}$	6
Combination — Lbs/Ft						27.4	31.6	33.3	36.5	43.4	49.5	55.8
Calcium Silicate — Thickness—In.	1½	1½	1½	1½	2	2½	3	3	3½	4	4½	5
Calcium Silicate — Lbs/Ft	5.91	5.91	5.91	5.91	7.92	10.1	12.3	12.3	14.7	17.2	19.8	22.5

FLANGES

Pressure Rating psi	Cast Iron 125	Cast Iron 250	Steel 150	Steel 300	Steel 400	Steel 600	Steel 900	Steel 1500	Steel 2500
Screwed or Slip-On	71 / 1.5	137 / 1.5	72 / 1.5	140 / 1.5	163 / 1.5	261 / 1.5	388 / 1.5	820 / 1.5	1611 / 1.5
Welding Neck			69 / 1.5	142 / 1.5	181 / 1.5	233 / 1.5	361 / 1.5	691 / 1.5	1671 / 1.5
Lap Joint			77 / 1.5	169 / 1.5	193 / 1.5	293 / 1.5	445 / 1.5	920 / 1.5	1591 / 1.5
Blind	96 / 1.5	177 / 1.5	118 / 1.5	209 / 1.5	260 / 1.5	341 / 1.5	475 / 1.5	928 / 1.5	1775 / 1.5

FLANGED FITTINGS

	125	250	150	300	400	600	900	1500	2500
S.R. 90° Elbow	265 / 5	453 / 5.2	345 / 5	509 / 5.2	669 / 5.5	815 / 5.8	1474 / 6.2		
L.R. 90° Elbow	375 / 6.2	553 / 6.2	485 / 6.2	624 / 6.2			1598 / 6.2		
45° Elbow	235 / 4.3	383 / 4.3	282 / 4.3	414 / 4.3	469 / 4.5	705 / 4.7	1124 / 4.8		
Tee	403 / 7.5	684 / 7.8	513 / 7.5	754 / 7.8	943 / 8.3	1361 / 8.7	1928 / 9.3		

**VALVES

	125	250	150	300	400	600	900	1500	2500
Flanged Bonnet Gate	687 / 7.8	1298 / 8.5	726 / 4.8	1337 / 5.5	1611 / 6.8	2283 / 7.1	3248 / 7.8		
Flanged Bonnet Globe or Angle	808 / 9.4	1200 / 9.5		1409 / 6.5	1493 / 6.8				
Flanged Bonnet Check	674 / 9.4	1160 / 9.5	701 / 6.5	874 / 6.5	1118 / 6.8	1168 / 7.1			
Pressure Seal Bonnet—Gate							2080 / 5.5	2400 / 5.9	
Pressure Seal Bonnet—Globe							2150 / 5.5	2500 / 5.9	

Bolts

	125	250	150	300	400	600	900	1500	2500
*One Complete Flanged Joint	15	44	15	49	69	91	124	306	622

SEE GENERAL NOTES FOR MATERIALS NOT SHOWN

All weights are shown in bold type.

The weight of steel pipe is per linear foot.

For Boiler Feed Piping, add the weight of water to the weight of steel pipe.

The pipe covering thicknesses and weights indicate the average conditions per linear foot and include all allowances for wire, cement, canvas, bands and paint. The listed thickness of combination covering is the sum of the inner and the outer layer thickness.

Pipe covering temperature ranges are intended as a guide only and do not constitute a recommendation for specific thickness of materials.

To find the weight of covering on Flanged Fittings, Valves, or Flanges multiply the weight factor (lightface subscript) by the weight per foot of covering used on straight pipe.

*All Flanged Fitting, Flanged Valve and Flange weights include the proportional weight of bolts or studs required to make up all joints.

**Cast Iron Valve weights are for flanged valves. Steel Valve weights are for welding end valves.

WEIGHTS OF PIPING MATERIALS—14" PIPE SIZE

PIPE

Schedule No.	10	20	30	40		60	80	100	120	140	160
Wall Designation			Std.		XS						
Thickness—In.	.250	.312	.375	.437	.500	.593	.750	.937	1.093	1.250	1.406
Pipe—Lbs/Ft	36.71	45.7	54.6	63.4	72.1	84.9	106.1	130.7	150.7	170.2	189.1
Water—Lbs/Ft	62.06	60.92	59.7	58.7	57.5	55.9	53.2	50.0	47.5	45.0	42.6

WELDING FITTINGS

Fitting	(Std.)	(XS)
L.R. 90° Elbow	154 / 3.5	202 / 3.5
S.R. 90° Elbow	102 / 2.3	135 / 2.3
L.R. 45° Elbow	77 / 1.5	100 / 1.5
Tee	159 / 2.8	203 / 2.8
Lateral	495 / 5.8	588 / 5.8
Reducer	63 / 1.1	83 / 1.1
Cap	35 / 1.7	46 / 1.7

COVERING

Temperature Range °F	to 260°	260-360	360-440	440-525	525-600	600-700	700-800	800-900	900-1000			
85% Magnesia Thickness—In.	1½	2	2½	3	3							
85% Magnesia Lbs/Ft	10.4	13.4	16.8	20.3	20.3							
Combination Thickness—In.						3½	4	4	4½	5	5½	6
Combination Lbs/Ft						28.7	33.1	33.1	39.4	46.4	53.0	59.7
Calcium Silicate Thickness—In.	1	1	1½	1½	2	2½	3	3½	4	4½	5	
Calcium Silicate Lbs/Ft	3.89	3.89	5.90	5.90	8.03	10.3	12.6	15.1	17.7	20.5	23.3	

FLANGES

Pressure Rating psi	Cast Iron 125	Cast Iron 250	Steel 150	Steel 300	Steel 400	Steel 600	Steel 900	Steel 1500	Steel 2500
Screwed or Slip-On	93 / 1.5	184 / 1.5	96 / 1.5	195 / 1.5	235 / 1.5	318 / 1.5	460 / 1.5		
Welding Neck			90 / 1.5	192 / 1.5	240 / 1.5	358 / 1.5	473 / 1.5		
Lap Joint			116 / 1.5	226 / 1.5	261 / 1.5	358 / 1.5	491 / 1.5		
Blind	126 / 1.5	239 / 1.5	142 / 1.5	267 / 1.5	354 / 1.5	437 / 1.5	648 / 1.5		

FLANGED FITTINGS

Fitting	125	250	150	300	400	600	900	1500	2500
S.R. 90° Elbow	372 / 5.3	617 / 5.5	497 / 5.3	632 / 5.5	664 / 5.7	918 / 5.9	1549 / 6.4		
L.R. 90° Elbow	492 / 6.6	767 / 6.6	622 / 6.6	772 / 6.6					
45° Elbow	292 / 4.3	497 / 4.4	377 / 4.3	587 / 4.4	638 / 4.6	883 / 4.8	1246 / 4.9		
Tee	563 / 8	956 / 8.4	683 / 8	968 / 8.3	1131 / 8.6	1652 / 8.9	2318 / 9.6		

VALVES

Valve	125	250	150	300	400	600	900	1500	2500
Flanged Bonnet Gate	921 / 7.9	1762 / 8.8	830 / 4.9	1872 / 6.3	2018 / 7.1	3082 / 7.4	3989 / 8.1		
Flanged Bonnet Globe or Angle	1171 / 9.9								
Flanged Bonnet Check	885 / 9.9								
Pressure Seal Bonnet—Gate									
Pressure Seal Bonnet—Globe									

Bolts

	125	250	150	300	400	600	900	1500	2500
*One Complete Flanged Joint	22	57	22	62	88	118	159		

SEE GENERAL NOTES FOR MATERIALS NOT SHOWN

All weights are shown in bold type.

The weight of steel pipe is per linear foot.

For Boiler Feed Piping, add the weight of water to the weight of steel pipe.

The pipe covering thicknesses and weights indicate the average conditions per linear foot and include all allowances for wire, cement, canvas, bands and paint. The listed thickness of combination covering is the sum of the inner and the outer layer thickness.

Pipe covering temperature ranges are intended as a guide only and do not constitute a recommendation for specific thickness of materials.

To find the weight of covering on Flanged Fittings, Valves, or Flanges, multiply the weight factor (lightface subscript) by the weight per foot of covering used on straight pipe.

*All Flanged Fitting, Flanged Valve and Flange weights include the proportional weight of bolts or studs required to make up all joints.

**Cast Iron Valve weights are for flanged valves. Steel Valve weights are for welding end valves.

WEIGHTS OF PIPING MATERIALS—16″ PIPE SIZE

PIPE

Schedule No.	10	20	30	40	60	80	100	120	140	160
Wall Designation			Std.	XS						
Thickness—In.	.250	.312	.375	.500	.565	.843	1.031	1.218	1.437	1.593
Pipe—Lbs/Ft	42.1	52.4	62.6	82.8	107.5	136.5	164.8	192.3	223.6	245.1
Water—Lbs/Ft	81.8	80.5	79.1	76.5	73.4	69.7	66.1	62.6	58.6	55.9

WELDING FITTINGS

	Sch. 30	Sch. 40
L.R. 90° Elbow	201 / 4	265 / 4
S.R. 90° Elbow	135 / 2.5	177 / 2.5
L.R. 45° Elbow	100 / 1.7	132 / 1.7
Tee	202 / 3.2	257 / 3.2
Lateral	650 / 6.7	774 / 6.7
Reducer	78 / 1.2	102 / 1.2
Cap	44 / 1.8	58 / 1.8

COVERING

Temperature Range °F	to 260°	260-360	360-440	440-525	525-600	600-700	700-800	800-900	900-1000			
85% Magnesia Thickness—In.	1½	2	2½	3	3							
85% Magnesia Lbs/Ft	11.6	15.1	18.8	22.6	22.6							
Combination Thickness—In.						3½	4	4	4½	5	5½	6
Combination Lbs/Ft						32.0	37.2	37.2	43.6	51.4	58.4	65.7
Calcium Silicate Thickness—In.	1	1	1½	1½	2	2½	3	3½	4	4½	5	
Calcium Silicate Lbs/Ft	4.40	4.40	6.65	6.65	9.02	11.5	14.1	16.8	19.7	22.7	25.8	

FLANGES

Pressure Rating psi	Cast Iron 125	Cast Iron 250	Steel 150	Steel 300	Steel 400	Steel 600	Steel 900	Steel 1500	Steel 2500
Screwed or Slip-On	120 / 1.5	233 / 1.5	108 / 1.5	261 / 1.5	310 / 1.5	442 / 1.5	559 / 1.5		
Welding Neck			116 / 1.5	257 / 1.5	308 / 1.5	492 / 1.5	564 / 1.5		
Lap Joint			151 / 1.5	289 / 1.5	347 / 1.5	489 / 1.5	607 / 1.5		
Blind	175 / 1.5	308 / 1.5	185 / 1.5	348 / 1.5	455 / 1.5	603 / 1.5	809 / 1.5		

FLANGED FITTINGS

	Cast Iron 125	Cast Iron 250	Steel 150	Steel 300	Steel 400	Steel 600	Steel 900	Steel 1500	Steel 2500
S.R. 90° Elbow	501 / 5.5	826 / 5.8	656 / 5.5	958 / 5.8	1014 / 6	1402 / 6.3	1886 / 6.7		
L.R. 90° Elbow	701 / 7	1036 / 7	781 / 7	1058 / 7					
45° Elbow	391 / 4.3	696 / 4.6	481 / 4.3	708 / 4.6	839 / 4.7	1212 / 5	1586 / 5		
Tee	746 / 8.3	1263 / 8.7	961 / 8.3	1404 / 8.6	1671 / 9	2128 / 9.4	3054 / 10		

VALVES

	Cast Iron 125	Cast Iron 250	Steel 150	Steel 300	Steel 400	Steel 600	Steel 900	Steel 1500	Steel 2500
Flanged Bonnet Gate	1254 / 8	2321 / 9	1315 / 5	2511 / 7.1	2694 / 7.5	3668 / 7.9			
Flanged Bonnet Globe or Angle									
Flanged Bonnet Check	1166 / 10.5								
Pressure Seal Bonnet—Gate									
Pressure Seal Bonnet—Globe									

Bolts

	Cast Iron 125	Cast Iron 250	Steel 150	Steel 300	Steel 400	Steel 600	Steel 900	Steel 1500	Steel 2500
*One Complete Flanged Joint	31	76	31	83	114	152	199		

SEE GENERAL NOTES FOR MATERIALS NOT SHOWN

All weights are shown in bold type.

The weight of steel pipe is per linear foot.

For Boiler Feed Piping, add the weight of water to the weight of steel pipe.

The pipe covering thicknesses and weights indicate the average conditions per linear foot and include all allowances for wire, cement, canvas, bands and paint. The listed thickness of combination covering is the sum of the inner and the outer layer thickness.

Pipe covering temperature ranges are intended as a guide only and do not constitute a recommendation for specific thickness of materials.

To find the weight of covering on Flanged Fittings, Valves, or Flanges, multiply the weight factor (lightface subscript) by the weight per foot of covering used on straight pipe.

*All Flanged Fitting, Flanged Valve and Flange weights include the proportional weight of bolts or studs required to make up all joints.

**Cast Iron Valve weights are for flanged valves. Steel Valve weights are for welding end valves.

WEIGHTS OF PIPING MATERIALS—18″ PIPE SIZE

	Schedule No.	10	20		30		40	60	80	100	120	140	160
PIPE	Wall Designation			Std.		XS							
	Thickness—In.	.250	.312	.375	.437	.500	.563	.750	.937	1.156	1.375	1.562	1.781
	Pipe—Lbs/Ft	47.4	59.0	70.6	82.1	93.5	104.8	138.2	170.8	208.0	244.1	274.2	308.5
	Water—Lbs/Ft	104.3	102.8	101.2	99.9	98.4	97.0	92.7	88.5	83.7	79.2	75.3	71.0

WELDING FITTINGS	L.R. 90° Elbow			256 / 4.5		338 / 4.5							
	S.R. 90° Elbow			171 / 2.8		225 / 2.8							
	L.R. 45° Elbow			128 / 1.9		168 / 1.9							
	Tee			258 / 3.6		328 / 3.6							
	Lateral			798 / 7.5		984 / 7.5							
	Reducer			94 / 1.3		123 / 1.3							
	Cap			57 / 2.1		75 / 2.1							

	Temperature Range °F	to 260°	260-360	360-440	440-525	525-600	600-700	700-800	800-900	900-1000			
COVERING	85% Magnesia — Thickness—In.	1½	2	2½	3	3							
	85% Magnesia — Lbs/Ft	12.9	16.0	20.8	25.0	25.0							
	Combination — Thickness—In.						3½	4	4	4½	5	5½	6
	Combination — Lbs/Ft						35.3	40.9	40.9	48.0	56.3	64.0	71.7
	Calcium Silicate — Thickness—In.	1	1	1½	1½	2	2½	3	3½	4	4½	5	
	Calcium Silicate — Lbs/Ft	4.91	4.91	7.40	7.40	10.0	12.7	15.6	18.6	21.7	25.9	28.2	

	Pressure Rating psi	Cast Iron		Steel							
		125	250	150	300	400	600	900	1500	2500	
FLANGES	Screwed or Slip-On	140 / 1.5		140 / 1.5	331 / 1.5	380 / 1.5	573 / 1.5	797 / 1.5			
	Welding Neck			128 / 1.5	316 / 1.5	377 / 1.5	569 / 1.5	786 / 1.5			
	Lap Joint			176 / 1.5	365 / 1.5	428 / 1.5	584 / 1.5	850 / 1.5			
	Blind	210 / 1.5	396 / 1.5	229 / 1.5	441 / 1.5	572 / 1.5	762 / 1.5	1152 / 1.5			
FLANGED FITTINGS	S.R. 90° Elbow	621 / 5.8	1060 / 6	711 / 5.8	1126 / 6	1340 / 6.2	1793 / 6.6	2817 / 7			
	L.R. 90° Elbow	881 / 7.4	1350 / 7.4	941 / 7.4	1426 / 7.4						
	45° Elbow	461 / 4.4	870 / 4.7	521 / 4.4	901 / 4.7	1040 / 4.8	1543 / 5	2252 / 5.2			
	Tee	921 / 8.6	1625 / 9	1010 / 8.6	1602 / 9	1909 / 9.3	2690 / 9.9	4327 / 10.5			
VALVES	Flanged Bonnet Gate	1629 / 8.2	2578 / 9.3		3189 / 7.5	3580 / 7.8	5647 / 8.4				
	Flanged Bonnet Globe or Angle										
	Flanged Bonnet Check	1371 / 10.5									
	Pressure Seal Bonnet—Gate										
	Pressure Seal Bonnet—Globe										
Bolts	*One Complete Flanged Joint	41	93	41	101	139	193	299			

SEE GENERAL NOTES FOR MATERIALS NOT SHOWN

All weights are shown in bold type.

The weight of steel pipe is per linear foot.

For Boiler Feed Piping, add the weight of water to the weight of steel pipe.

The pipe covering thicknesses and weights indicate the average conditions per linear foot and include all allowances for wire, cement, canvas, bands and paint. The listed thickness of combination covering is the sum of the inner and the outer layer thickness.

Pipe covering temperature ranges are intended as a guide only and do not constitute a recommendation for specific thickness of materials.

To find the weight of covering on Flanged Fittings, Valves, or Flanges, multiply the weight factor (lightface subscript) by the weight per foot of covering used on straight pipe.

*All Flanged Fitting, Flanged Valve and Flange weights include the proportional weight of bolts or studs required to make up all joints.

**Cast Iron Valve weights are for flanged valves. Steel Valve weights are for welding end valves.

WEIGHTS OF PIPING MATERIALS—20″ PIPE SIZE

PIPE	Schedule No.	10	20	30	40	60	80	100	120	140	160		
	Wall Designation		Std.	XS									
	Thickness—In.	.250	.375	.500	.593	.812	1.031	1.281	1.500	1.750	1.968		
	Pipe—Lbs/Ft	52.7	78.6	104.1	122.9	166.4	208.9	256.1	296.4	341.1	379.0		
	Water—Lbs/Ft	129.5	126.0	122.8	120.4	115.0	109.4	103.4	98.3	92.6	87.9		

WELDING FITTINGS

	20	30
L.R. 90° Elbow	317 / 5	419 / 5
S.R. 90° Elbow	212 / 3.4	278 / 3.4
L.R. 45° Elbow	158 / 2.1	208 / 2.1
Tee	321 / 4	407 / 4
Lateral	1024 / 8.3	1221 / 8.3
Reducer	142 / 1.7	186 / 1.7
Cap	72 / 2.3	94 / 2.3

COVERING

Temperature Range °F	to 260°	260-360	360-440	440-525	525-600	600-700	700-800	800-900	900-1000		
85% Magnesia Thickness—In.	1½	2	2½	3	3						
85% Magnesia Lbs/Ft	14.2	18.4	22.8	27.2	27.2						
Combination Thickness—In.						3½	4	4	4½	5	5½	6
Combination Lbs/Ft						38.6	44.7	44.7	52.2	61.2	69.4	77.8
Calcium Silicate Thickness—In.	1	1	1½	1½	2	2½	3	3½	4	4½	5	
Calcium Silicate Lbs/Ft	5.39	5.39	8.15	8.15	11.0	14.0	17.1	20.3	23.6	27.0	30.6	

FLANGES

	Cast Iron		Steel							
Pressure Rating psi	125	250	150	300	400	600	900	1500	2500	
Screwed or Slip-On	176 / 1.5		181 / 1.5	378 / 1.5	468 / 1.5	733 / 1.5	973 / 1.5			
Welding Neck			159 / 1.5	389 / 1.5	475 / 1.5	704 / 1.5	952 / 1.5			
Lap Joint			222 / 1.5	438 / 1.5	524 / 1.5	748 / 1.5	1085 / 1.5			
Blind	276 / 1.5	487 / 1.5	298 / 1.5	545 / 1.5	711 / 1.5	976 / 1.5	1438 / 1.5			

FLANGED FITTINGS

	125	250	150	300	400	600	900
S.R. 90° Elbow	792 / 6	1315 / 6.3	922 / 6	1375 / 6.3	1680 / 6.5	2314 / 6.9	3610 / 7.3
L.R. 90° Elbow	1132 / 7.8	1725 / 7.8	1352 / 7.8	1705 / 7.8			
45° Elbow	592 / 4.6	1055 / 4.8	652 / 4.6	1105 / 4.8	1330 / 4.9	1917 / 5.2	2848 / 5.4
Tee	1178 / 9	2022 / 9.5	1378 / 9	1908 / 9.5	2370 / 9.7	3463 / 10.1	5520 / 11

VALVES

	125	250	150	300	400	600	900
Flanged Bonnet Gate	1934 / 8.3	3823 / 9.5		4449 / 7.9	4744 / 8.2	6476 / 8.9	
Flanged Bonnet Globe or Angle							
Flanged Bonnet Check	1772 / 11						
Pressure Seal Bonnet—Gate							
Pressure Seal Bonnet—Globe							

Bolts

	125	250	150	300	400	600	900
*One Complete Flanged Joint	52	95	52	105	180	242	361

SEE GENERAL NOTES FOR MATERIALS NOT SHOWN

All weights are shown in bold type.

The weight of steel pipe is per linear foot.

For Boiler Feed Piping, add the weight of water to the weight of steel pipe.

The pipe covering thicknesses and weights indicate the average conditions per linear foot and include all allowances for wire, cement, canvas, bands and paint. The listed thickness of combination covering is the sum of the inner and the outer layer thickness.

Pipe covering temperature ranges are intended as a guide only and do not constitute a recommendation for specific thickness of materials.

To find the weight of covering on Flanged Fittings, Valves, or Flanges, multiply the weight factor (lightface subscript) by the weight per foot of covering used on straight pipe.

*All Flanged Fitting, Flanged Valve and Flange weights include the proportional weight of bolts or studs required to make up all joints.

**Cast Iron Valve weights are for flanged valves. Steel Valve weights are for welding end valves.

WEIGHTS OF PIPING MATERIALS—24″ PIPE SIZE

PIPE

	Schedule No.	10	20		30	40	60	80	100	120	140	160
	Wall Designation		Std.	XS								
	Thickness—In.	.250	.375	.500	.562	.687	.968	1.218	1.531	1.812	2.062	2.343
	Pipe—Lbs/Ft	63.4	94.6	125.5	140.8	171.2	238.1	296.4	367.4	429	484	541
	Water—Lbs/Ft	188.0	183.8	180.1	178.1	174.3	165.8	158.3	149.3	141	134	127

WELDING FITTINGS

Fitting	Std.	XS
L.R. 90° Elbow	458 / 6	606 / 6
S.R. 90° Elbow	305 / 3.7	404 / 3.7
L.R. 45° Elbow	229 / 2.5	302 / 2.5
Tee	445 / 4.9	563 / 4.9
Lateral	1482 / 10	1769 / 10
Reducer	167 / 1.7	220 / 1.7
Cap	102 / 2.8	134 / 2.8

COVERING

	Temperature Range °F	to 260°	260–360	360–440	440–525	525–600	600–700	700–800	800–900	900–1000				
85% Magnesia	Thickness—In.	1½	2	2½	3	3								
	Lbs/Ft	16.7	21.8	26.8	32.0	32.0								
Combination	Thickness—In.						3½	4	4	4½	5	5½	6	
	Lbs/Ft						45.2	52.2	52.2	60.8	71.2	80.4	90.0	
Calcium Silicate	Thickness—In.	1	1	1¼	1½	2	2½	3	4	4½	5			
	Lbs/Ft	6.44	6.44	9.65	9.65	13.0	16.4	20.0	27.5	31.4	35.5			

FLANGES

	Pressure Rating psi	Cast Iron 125	Cast Iron 250	Steel 150	300	400	600	900	1500	2500
	Screwed or Slip-On	255 / 1.5		245 / 1.5	577 / 1.5	676 / 1.5	1055 / 1.5	1823 / 1.5		
	Welding Neck			248 / 1.5	580 / 1.5	702 / 1.5	998 / 1.5	1793 / 1.5		
	Lap Joint			309 / 1.5	631 / 1.5	770 / 1.5	1080 / 1.5	2058 / 1.5		
	Blind	405 / 1.5	757 / 1.5	446 / 1.5	841 / 1.5	1073 / 1.5	1354 / 1.5	2715 / 1.5		

FLANGED FITTINGS

Fitting	125	250	150	300	400	600	900	1500	2500
S.R. 90° Elbow	1231 / 6.7	2014 / 6.8	1671 / 6.7	2174 / 6.8	2474 / 7.1	3506 / 7.6	6155 / 8.1		
L.R. 90° Elbow	1711 / 8.7	2644 / 8.7	1821 / 8.7	2874 / 8.7					
45° Elbow	871 / 4.8	1604 / 5	1121 / 4.8	1634 / 5	1974 / 5.1	2831 / 5.5	5124 / 6		
Tee	1836 / 10	3061 / 10.2	2276 / 10	3161 / 10.2	3811 / 10.6	5184 / 11.4	9387 / 12.1		

VALVES

Valve	125	250	150	300	400	600	900	1500	2500
Flanged Bonnet Gate	3062 / 8.5	6484 / 9.8		6920 / 8.7	7122 / 9.1	9246 / 9.9			
Flanged Bonnet Globe or Angle									
Flanged Bonnet Check	2956 / 12								
Pressure Seal Bonnet—Gate									
Pressure Seal Bonnet—Globe									

Bolts

	125	250	150	300	400	600	900
*One Complete Flanged Joint	71	174	71	174	274	360	687

SEE GENERAL NOTES FOR MATERIALS NOT SHOWN

All weights are shown in bold type.

The weight of steel pipe is per linear foot.

For Boiler Feed Piping, add the weight of water to the weight of steel pipe.

The pipe covering thicknesses and weights indicate the average conditions per linear foot and include all allowances for wire, cement, canvas, bands and paint. The listed thickness of combination covering is the sum of the inner and the outer layer thickness.

Pipe covering temperature ranges are intended as a guide only and do not constitute a recommendation for specific thickness of materials.

To find the weight of covering on Flanged Fittings, Valves, or Flanges, multiply the weight factor (lightface subscript) by the weight per foot of covering used on straight pipe.

*All Flanged Fitting, Flanged Valve and Flange weights include the proportional weight of bolts or studs required to make up all joints.

**Cast Iron Valve weights are for flanged valves. Steel Valve weights are for welding end valves.

HANGER LOAD CALCULATIONS

The thermal deflection of piping in modern high pressure and high temperature power stations has made it necessary to specify flexible supports, thereby requiring the designer to give considerable thought to the calculations of hanger loads for the steam pipe lines. Most manufacturers of power station equipment place limitations on the allowable loads at terminal points. Boiler and turbine manufacturers are especially concerned about the pipe weight on their equipment and sometimes specify that the reactions at pipe connections shall be zero. Piping engineers are interested in knowing the proper procedure to follow in calculating the loads at various points of support on a pipe line and also to determine the position of support on a pipe line where the pipe load at the anchor or terminal point is limited.

Pipe lines should be supported so that the load on the terminal points is not greater than the allowable load throughout the full range of thermal expansion.

Therefore, it is desirable to know the supporting force at each pipe suspension point and to have the total supporting forces equal to the calculated weight of the piping system.

With these conditions in mind, it is quite evident that the pipe support engineer has need for a method of calculating the supports that will be clear, concise and easily understood.

In the example problems to follow it is assumed that the hangers have been located. The approach and the assumptions made in solving a hanger load problem could be numerous, depending on the designer. The solution that follows is not intended as being the only method but rather as a method of producing a good balanced system for the problem under consideration. Of the approaches that could be made in the solution of any problem there will be one solution that produces the best balanced system. Individual loads will very likely be different but the total of each combination of hanger loads plus reactions obtained should be approximately the same in every case.

A well balanced suspension system will result in values for the loads on the hangers to be in close proximity to one another provided all pipe is of the same size or there are no highly concentrated loads located near a hanger; where there are concentrated

loads within the system the supporting forces required of the adjacent hangers will be correspondingly large.

Figure 1 illustrates a pipe line drawn in isometric with all the necessary dimensions shown in the same plane as the related section of pipe. This illustration is limited to as few pipe sections as possible but incorporates most of the problems commonly encountered in power station piping support. The type of support, spring or rigid, is not covered here as this is a function of the piping system's thermal deflection and should be treated as a separate study.

The first step in the solution of a problem of this kind is to prepare a table of weights. Such a table, for the pipe line shown in Figure 1, is given on the following page.

Calculation of loads is accomplished by taking moments about an unknown value and solving for a second unknown value, or if all loads except one are known, summation of the individual loads will produce the unknown load. It is preferred in this explanation that the summation method be used as check method on the accuracy of the computations made in the moment method.

The calculation of loads for hangers of a piping system involves first dividing the system into convenient sections at the following points in the system and the order of preference is as listed: (1) Hangers, (2) Bends (either vertical or horizontal), and (3) Risers.

The next step is to isolate each section for study beginning with the section of pipe supported by hangers H-1 and H-2. The solution of each section should be prefaced by the drawing of a free body diagram. This will provide a clearer picture of the steps involved in the solution.

Section 1

Draw a single line sketch, preferably to scale, as in Figure 2, and show all dimensions and weights. (The weight should be shown at the center of gravity of each piece of pipe, valve or fitting.)

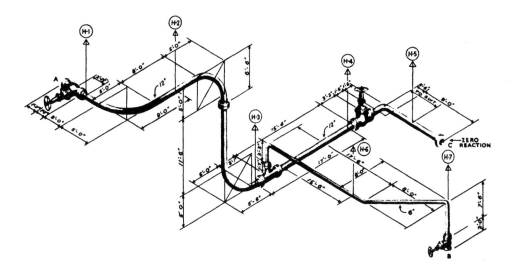

Figure 1

NOTE: Maximum Allowable Load at B = 650 lb

Table of Weights

Description	Weight in Pounds	Insulation Weight in Pounds	Total Weight	Weight Used in Calculations	Unit
12″-Schedule 160 Pipe	161	36.5	197.5	197.5	Per Ft
6″-Schedule 160 Pipe	45.3	24.2	69.5	69.5	Per Ft
12″-900 lb Stop and Check Valve	3960	394.0	4354.0	4354.0	Each
12″-900 lb Lap Flange	445	54.75	499.75	500.0	Each
12″ × 12″ × 6″-900 lb Flanged Tee	1683	321.0	2004.0	2004.0	Each
12″-900 lb Flanged Gate Valve	4024	394.0	4418.0	4418.0	Each
12″-Schedule 160 Weld Ell	460	109.5	569.5	570.0	Each
6″-W.N.F.S. 900 lb Flange	116	36.3	152.3	152.0	Each
6″-Schedule 160 Weld Ell	59	36.3	95.3	95.0	Each
6″-900 lb Flanged Globe Valve	1100	217.8	1317.8	1318.0	Each
12″-900 lb Flanged Elbow	1598	226.3	1824.3	1824.0	Each

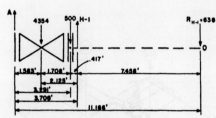

Figure 4

or

$(H\text{-}2)''$
$$= 99 + 1550 + 3173 + 1550 + 790 + 477 - 3804$$
$$= 3835\text{lb}$$

Total load on $H\text{-}2 = (H\text{-}2)' + (H\text{-}2)'' = 927 + 3835 = 4762\text{lb}.$

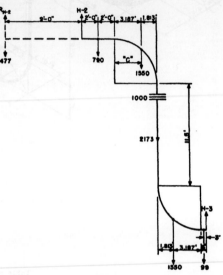

Figure 5

force on $H\text{-}1$ due to the line $A\text{-}(H\text{-}1)$ plus $R_{H\text{-}1}$ reacting at 0.

$3.708(H\text{-}1)'' = 4354(1.583) + 500(3.291) + 638(11.166)$
$\qquad = 6892.382 + 1645.5 + 7123.908$
$\qquad = 15,661.79$
$\quad (H\text{-}1)'' = 4224\text{lb}$

or

$\quad (H\text{-}1)'' = 4354 + 500 + 638 - 1268 = 4224\text{lb}$

Total load on $H\text{-}1 = (H\text{-}1)' + (H\text{-}1)'' = 841 + 4224 = 5065\text{lb}$

Section 2

Consider next the section of pipe between $H\text{-}2$ and $H\text{-}3$. Figure 5 shows the section in elevation with the loads indicated as in Figure 2. In this section we will consider $R_{H\text{-}2}$, which as yet has not been balanced.

The weight of the vertical bend is considered as acting at the center of gravity of the bend. Figure 3 can be used to determine this location.

$$C = 5 \times 0.637 = 3.183' = 3'\text{-}2\tfrac{1}{4}''$$

All forces are in the vertical plane.

Take moments about $H\text{-}2$, solve for $(H\text{-}3)'$, the load on $H\text{-}3$ due to the line $(H\text{-}2)\text{-}(H\text{-}3)$.

$14.5(H\text{-}3)' = -477(9) + 790(2) + 1550(7.187)$
$\qquad + 3173(9) + 1550(10.183) + 99(14.25)$
$\qquad = -4293 + 1580 + 11,139.85 + 28,557$
$\qquad + 16,760.15 + 1410.75$
$\qquad = 55,154.75$
$\quad (H\text{-}3)' = 3804\text{lb}$

Take moments about $H\text{-}3$ to solve for $(H\text{-}2)''$, the load on $H\text{-}2$ due to line $(H\text{-}2)$ to $(H\text{-}3)$.

$14.5(H\text{-}2)'' = 99(0.25) + 1550(3.683) + 3173(5.5)$
$\qquad + 1550(7.317) + 790(12.5)$
$\qquad + 477(23.5)$
$\qquad = 24.75 + 5708.65 + 17,451.5$
$\qquad + 11,341.35 + 9875 + 11,209.5$
$\qquad = 55,610.75$
$\quad (H\text{-}2)'' = 3835\text{lb}$

Section 3

The section of pipe between $H\text{-}4$ and $H\text{-}5$ has an imposed load of the 6″ line through the flanged tee. This load must now be determined and at the same time solve for loads on Hangers $H\text{-}6$ and $H\text{-}7$. See Figure 6.

The load and the imaginary beam reactions for the 45° bend are calculated as in Section 1. The load due to line $(H\text{-}6)\text{-}R_{H\text{-}7}$ results in the reaction $R_{H\text{-}7}$ which is to be carried by $H\text{-}7$. The load due to line $R_{H\text{-}6}\text{-}(H\text{-}7)$ results in the reaction $R_{H\text{-}6}$ which is to be carried by $H\text{-}6$.

Taking moments about $H\text{-}6$ solve for $R_{H\text{-}7}$, see Figure 7.

$2R_{H\text{-}7} = 62(0.437) + 146(1.208)$
$\qquad = 27.094 + 176.368$
$\qquad = 203.462$
$\quad R_{H\text{-}7} = 102\text{lb}$

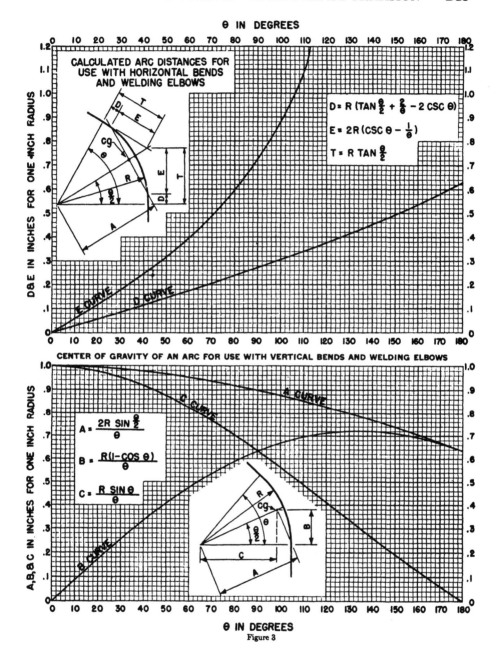

θ IN DEGREES

CALCULATED ARC DISTANCES FOR USE WITH HORIZONTAL BENDS AND WELDING ELBOWS

$$D = R \left(TAN \frac{\theta}{2} + \frac{2}{\theta} - 2 \, CSC \, \theta \right)$$

$$E = 2R \left(CSC \, \theta - \frac{1}{\theta} \right)$$

$$T = R \, TAN \frac{\theta}{2}$$

D&E IN INCHES FOR ONE INCH RADIUS

E CURVE

D CURVE

CENTER OF GRAVITY OF AN ARC FOR USE WITH VERTICAL BENDS AND WELDING ELBOWS

A, B & C IN INCHES FOR ONE INCH RADIUS

$$A = \frac{2R \, SIN \frac{\theta}{2}}{\theta}$$

$$B = \frac{R(1 - COS \, \theta)}{\theta}$$

$$C = \frac{R \, SIN \, \theta}{\theta}$$

A CURVE

C CURVE

B CURVE

θ IN DEGREES

Figure 3

The weight of the 90° bend is shown as 1550 lb at the center of gravity of the bend. Consider this bend as supported on a beam which passes through the center of gravity and rests on the extensions of the tangents to the bend. This imaginary beam is shown resting on the tangents at a distance D of $1'\text{-}4\frac{1}{2}''$ and the load on each end of the beam is one half the total load or 775 lb.

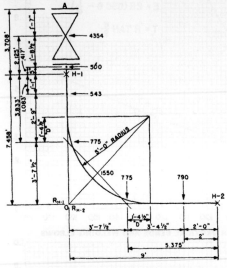

Figure 2.   Horizontal Bend—Plan View

The distance D is determined trigonometrically or from Figure 3 which has been drawn for convenience.

$$D = 5 \times 0.273 = 1.365' \quad \text{or} \quad 1'\text{-}4\frac{1}{2}''$$

Now consider the forces between H-1 and H-2 acting in two planes which intersect at 0. There will be two reactions at 0 which are designated as $R_{H\text{-}1}$ and $R_{H\text{-}2}$. $R_{H\text{-}1}$ is the reaction of line 0–(H-2) to be carried on Hanger H-1 and $R_{H\text{-}2}$ is the reaction of line (H-1)–0 to be carried on Hanger H-2. Transpose the feet and inches to their decimal equivalent in feet. Consider line (H-1)–0 as a free body and by taking moments about H-1 solve for $R_{H\text{-}2}$.

$$7.458 R_{H\text{-}2} = 543(1.083) + 775(3.833)$$
$$= 588.069 + 2970.575$$
$$= 3558.664$$
$$R_{H\text{-}2} = 477\,\text{lb}$$

By taking moments about $R_{H\text{-}2}$ solve for $(H\text{-}1)'$, the

force on H-1 due to the line (H-1)–0.

$$7.458(H\text{-}1)' = 775(3.625) + 543(6.375)$$
$$= 2809.375 + 3460.625$$
$$= 6270$$
$$(H\text{-}1)' = 841\,\text{lb}$$

or

$$(H\text{-}1)' = 543 + 775 - 477 = 841\,\text{lb}$$

This latter method of calculating $(H\text{-}1)'$ can be used as a check on the work of calculating the loads by taking moments, it consists of the sum of the loads minus the reaction.

Consider line 0–(H-2) as a free body and by taking moments about H-2 solve for $R_{H\text{-}1}$.

$$9 R_{H\text{-}1} = 790(2) + 775(5.375)$$
$$= 1580 + 4165.625$$
$$= 5745.625$$
$$R_{H\text{-}1} = 638\,\text{lb}$$

By taking moments about $R_{H\text{-}1}$ solve for $(H\text{-}2)'$, the force on H-2 due to the line 0–(H-2).

$$9(H\text{-}2)' = 775(3.625) + 790(7)$$
$$= 2809.375 + 5530$$
$$= 8339.375$$
$$(H\text{-}2)' = 927\,\text{lb}$$

or

$$(H\text{-}2)' = 775 + 790 - 638 = 927\,\text{lb}$$

For the section of pipe considered, H-1 to H-2, we have reactions as follows:

$$(H\text{-}1)' = 841\,\text{lb}$$
$$(H\text{-}2)' = 927\,\text{lb}$$
$$R_{H\text{-}1} = 638\,\text{lb}$$
$$R_{H\text{-}2} = 477\,\text{lb}$$

We must now determine the load on H-1 due to the forces between H-1 and A. By definition we said that the force resulting from (H-1)–0 was to be carried by H-2, this means that this section is to be balanced by the section between H-2 and H-3 so that in calculating the load on A, section (H-1)–0 is considered weightless.

Conversely, section 0–(H-2), which results in reaction $R_{H\text{-}1}$ at 0, is to be carried by H-1 and therefore balanced by section A–(H-1). Section A–0 in this case is a simple beam and is solved by taking moments about H-1 to find the reaction at A.

$$3.708A = 4354(2.125) + 500(0.417) - 638(7.458)$$
$$= 9252.25 + 208.5 - 4758.204$$
$$= 4702.546$$
$$A = 1268\,\text{lb} \text{ which is below the allowable load at } A \text{ of } 1500\,\text{lb}$$

Taking moments about A solve for $(H\text{-}1)''$, the

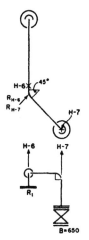

Figure 6

$$11.3R_{H-6} = 95(0.25) + 653(5.458) + 146(10.5)$$
$$= 23.75 + 3564.074 + 1533$$
$$= 5120.824$$
$$R_{H-6} = 453\text{lb}$$

Taking moments about R_{H-6} solve for $(H-7)'$, the load on $H-7$ due to the line $R_{H-6}-(H-7)$.

$$11.3(H-7)' = 146(0.800) + 653(5.834) + 95(11.05)$$
$$= 116.8 + 3809.602 + 1049.75$$
$$= 4976.152$$
$$(H-7)' = 441\text{lb}$$

or

$$(H-7)' = 146 + 653 + 95 - 453 = 441\text{lb}$$

Next consider the section between $H-7$ and B plus the reaction R_{H-7}, see Figure 9.

Taking moments about R_{H-7} solve for $(H-6)'$, the load on $H-6$ due to the line $(H-6)-R_{H-7}$.

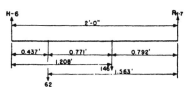

Figure 7

$$2(H-6)' = 146(0.792) + 62(1.563)$$
$$= 115.632 + 96.906$$
$$= 212.538$$
$$(H-6)' = 106\text{lb}$$

or

$$(H-6)' = 146 + 62 - 102 = 106\text{lb}$$

Figure 8 is a free body diagram of section $R_{H-6}-(H-7)$. Take moments about $H-7$ to solve for R_{H-6}.

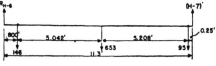

Figure 8

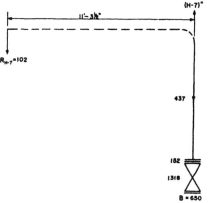

Figure 9

It is necessary to treat this section as a cantilever beam. The load on $H-7$ caused by this section is $(H-7)''$ and can be solved for by summing the forces.

$$(H-7)'' = 102 + 437 + 152 + 1318 - 650 = 1359\text{lb}$$

Total load on $H-7 = (H-7)' + (H-7)'' = 441 + 1359 = 1800\text{lb}$.

Section 4

Figure 10 is an elevation view of the section of 6″ pipe between $H-6$ and the flanged tee with the reaction R_{H-6}, which is the load on $H-6$ due to the line $R_{H-6}-(H-7)$.

Taking moments about R_1 solve for $(H\text{-}6)''$.

$$15.5(H\text{-}6)'' = 95(0.25) + 1025(8.125) + 453(17.5)$$
$$= 23.75 + 8328.125 + 7927.5$$
$$= 16,279.375$$
$$(H\text{-}6)'' = 1050 \text{lb}$$

Total load on $H\text{-}6 = (H\text{-}6)' + (H\text{-}6)'' = 106 + 1050 = 1156 \text{lb}$.

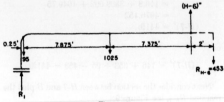

Figure 10

Taking moments about $(H\text{-}6)''$ solve for R_1, the load on the flanged tee.

$$15.5R_1 = 1025(7.375) + 95(15.25) + 140(15.5) - 453(2)$$
$$= 7559.375 + 1448.75 + 2170 - 906$$
$$= 10,272.125$$
$$R_1 = 663 \text{lb}$$

or

$$R_1 = 140 + 95 + 1025 + 453 - 1050 = 663 \text{lb}$$

Section 5

Figure 11 shows the pipe section $(H\text{-}3)$-$(H\text{-}4)$ as a simple beam. Solve for the reaction $(H\text{-}3)''$ by taking moments about $H\text{-}4$.

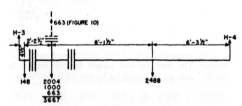

Figure 11

$$17(H\text{-}3)'' = 2488(6.292) + 3667(14.417) + 148(16.625)$$
$$= 15,654.496 + 52,867.139 + 2460.5$$
$$= 70,982.135$$
$$(H\text{-}3)'' = 4175 \text{lb}$$

Total load on $H\text{-}3 = (H\text{-}3)' + (H\text{-}3)'' = 3804 + 4175 = 7979 \text{lb}$.

Take moments about $H\text{-}3$ and solve for $(H\text{-}4)'$, the load on $H\text{-}4$ due to the $H\text{-}3$-$H\text{-}4$ line.

$$17(H\text{-}4)' = 148(0.375) + 3667(2.583) + 2488(10.708)$$
$$= 55.5 + 9472.861 + 26,641.504$$
$$= 36,169.865$$
$$(H\text{-}4)' = 2128 \text{lb}$$

or

$$(H\text{-}4)' = 148 + 3667 + 2488 - 4175 = 2128 \text{lb}$$

Section 6

Referring to Figure 1 it will be noted that it is required that there be zero reaction at point marked C.

Draw a sketch to scale as in Figure 12 showing loads and dimensions.

Consider the forces as acting in two planes which intersect at 0.

Calculate the load $(H\text{-}4)''$ on $H\text{-}4$ due to the section $(H\text{-}4)$ to 0 by taking moments about 0.

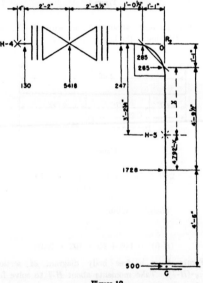

Figure 12

$$7.083(H\text{-}4)'' = 285(1.083) + 247(2.125)$$
$$+ 5418(4.583) + 130(6.75)$$
$$= 308.655 + 524.875 + 24,830.694 + 877.5$$
$$= 26,541.724$$
$$(H\text{-}4)'' = 3747 \text{lb}$$

Total load on $H\text{-}4 = (H\text{-}4)' + (H\text{-}4)'' = 2128 + 3747. = 5875 \text{lb}$.

Calculate the reaction at R_2 due to loads in the plane H-4 to 0 by taking moments about H-4.

$$7.083R_2 = 130(0.333) + 5418(2.5) + 247(4.958) + 285(6)$$
$$= 43.29 + 13,545 + 1224.626 + 1710$$
$$= 16,522.916$$
$$R_2 = 2333 \text{lb}$$

or

$$R_2 = 130 + 5418 + 247 + 285 - 3747 = 2333 \text{lb}$$

Solve for load and location of H-5 by taking moments about H-5.

$$2333(1.083 + X) + 285(X)$$
$$= 1728(4.792 - X) + 500(4.5 + 4.792 - X)$$
$$2526.639 + 2333X + 285X$$
$$= 8280.576 - 1728X + 2250 + 2396 - 500X$$
$$2333X + 285X + 1728X + 500X$$
$$= 8280.576 + 2250 + 2396 - 2526.639$$
$$4846X = 10,399.937$$
$$X = 2.146' = 2' \text{-} 1\tfrac{3}{4}''$$

Location of H-5 is $2'\text{-}1\tfrac{3}{4}'' + 1'\text{-}1''$ or $3'\text{-}2\tfrac{3}{4}''$ from 0.
Check

$$2333(3.229) = 7533.257$$

$$285(2.146) = \underline{611.610}$$
$$8144.867 \text{ ft lb}$$

$$1728(2.646) = 4572.288$$
$$500(7.146) = \underline{3573.000}$$
$$8145.288 \text{ ft lb}$$

Take moments about R_2 to solve for H-5.

$$3.229 H\text{-}5 = 285(1.083) + 1728(5.875) + 500(10.375)$$
$$= 308.655 + 10,152 + 5187.5$$
$$= 15,648.155$$
$$H\text{-}5 = 4846 \text{lb}$$

or

$$H\text{-}5 = 2333 + 285 + 1728 + 500 = 4846 \text{lb}$$

Summary:
The loads to be supported by each of the seven hangers as determined in the foregoing calculations are as follows:

$$H\text{-}1 = 5065 \text{lb}$$
$$H\text{-}2 = 4762 \text{lb}$$
$$H\text{-}3 = 7979 \text{lb}$$
$$H\text{-}4 = 5875 \text{lb}$$
$$H\text{-}5 = 4846 \text{lb}$$
$$H\text{-}6 = 1156 \text{lb}$$
$$H\text{-}7 = 1800 \text{lb}$$

MINUTES TO DECIMAL HOURS
CONVERSION TABLE

Minutes	Hours	Minutes	Hours
1	.017	31	.517
2	.034	32	.534
3	.050	33	.550
4	.067	34	.567
5	.084	35	.584
6	.100	36	.600
7	.117	37	.617
8	.135	38	.634
9	.150	39	.650
10	.167	40	.667
11	.184	41	.684
12	.200	42	.700
13	.217	43	.717
14	.232	44	.734
15	.250	45	.750
16	.267	46	.767
17	.284	47	.784
18	.300	48	.800
19	.317	49	.817
20	.334	50	.834
21	.350	51	.850
22	.368	52	.867
23	.384	53	.884
24	.400	54	.900
25	.417	55	.917
26	.434	56	.934
27	.450	57	.950
28	.467	58	.967
29	.484	59	.984
30	.500	60	1.000

Printed and bound by CPI Group (UK) Ltd, Croydon, CR0 4YY

08/05/2025

01864834-0003